A Guide to Empirical Orthogonal Functions for Climate Data Analysis

Antonio Navarra · Valeria Simoncini

A Guide to Empirical Orthogonal Functions for Climate Data Analysis

Springer

Dr. Antonio Navarra
Ist. Nazionale di Geofisica e
Vulcanologia
Via Gobetti, 101
40100 Bologna
Italy
navarra@ingv.it

Prof. Valeria Simoncini
Università di Bologna
Dip. to Matematica
Piazza di Porta San Donato, 5
40126 Bologna
Italy

Additional material to this book can be downloaded from http://extra.springer.com.

ISBN 978-94-007-9160-2 ISBN 978-90-481-3702-2 (eBook)
DOI 10.1007/978-90-481-3702-2
Springer Dordrecht Heidelberg London New York

Cover design: Boekhorst Design b.v.

Printed on acid-free paper

Springer is part of Springer Science+Business Media (www.springer.com)

Contents

Chapter 1
Introduction

Climatology and meteorology has been basically a descriptive science without the means to perform quantitative experiments under controlled conditions. In fact, until the second half of the twentieth century, the border between climatology and geography was often blurred and the two disciplines were confused one with the other.The situation changed when the solution of the evolution equations for the climate system became possible using numerical methods. The development of numerical models allowed the application of standard scientific verification machinery for testing hypotheses, but crucial to the success of the strategy is that the model must be a good representation of the real climate system of the Earth. Assessing the quality of models regarding their capability to reproduce the climate became a cornerstone in the scientific progress of climatology. Tighter and tighter standards were required for the model simulations in comparison with the real characteristics of climate. Models were required to reproduce not only the mean properties of climate, but also its variability. In the last decades of the XX century the amount of data available was becoming very large and strong evidence of remote spatial relations between climate variability in geographically diverse regions were emerging. Quantitative techniques were developed to explore the climate variability and its relations among different geographical locations. Methods were borrowed from descriptive statistics, where they were developed to analyze variance of related observations-variable pairs, or to identify unknown relations among variables.

These methods were introduced to meteorology in the mid-1960, but they became increasingly popular in the early 1980s where their capability to identify dynamically significant modes in the climate variability was demonstrated. Since then they have been further developed and many variants and extensions have been proposed and applied. Very often these developments were taking place separately from the formal development in the mainstream statistics and reflected ad hoc solution to the particular vies that climatology was using.

There are excellent books treating these methods in a formal and rigorous way (von Storch and Zwiers 1999; Wilks 2005; Jolliffe 2002) and we refer the reader to these excellent texts for proofs and a more formal treatment. We take in this booklet a different approach, trying to introduce the reader to a practical application of the methods and to the kind of real problems that can be encountered in a practical application. We are including in the book data sets from real simulations

A. Navarra and V. Simoncini, *A Guide to Empirical Orthogonal Functions for Climate Data Analysis*, DOI 10.1007/978-90-481-3702-2_1,

from a climate model that we use as a workbench to illustrate the various methods, their limitations and their potential. The algorithms have been implemented using MATLAB and we include some sample source codes for a few of the key methods employed. All pictures and examples used in the book may be reproduced by using the data sets and the routines available in the book WebSite.

Though the main thrust of the book is for climatological and meteorological examples, we feel that the treatment is sufficiently general that the discussion is useful also for student and practitioners in other fields, as long as they deal with variables that depend on two parameters like for instance space and time. This book can then be used as a reference for practical applications or it can be used in graduate courses as a companion to a more extensive book like (von Storch and Zwiers 1999; Wilks 2005; Preisendorfer 1988).

For most of this book we will use two data sets to illustrate our discussion. The first is a time series of monthly mean geopotential data at 500 mb (Z500), obtained from a simulation with a general circulation model forced by observed values of monthly mean Sea Surface Temperatures (SST). The data sets cover 34 years, corresponding to the calendar years 1961–1994. Z500 is a very good indicator of upper air flow, since the horizontal wind is predominantly aligned along the geopotential isolines. Figure 4.1 shows a few examples taken from the data set. It is possible to note the large variability from 1 month to the other (top panels), but also the large variability at the same geographical point, as the time series for the entire series (lower panels) show. It is clear that the geopotential at 500 mb is characterized by intense variability in space and time and a typical month may be as different from the next month as another one chosen at random. The large variability in space and time makes it a very good test case to practice at will. In the final chapter we will use two more time series, also 34 years long, obtained imposing the same SST distribution but with a small perturbation in the initial condition of the atmospheric numerical model. These small errors grow very quickly in a typical expression of the chaotic nature of the atmosphere and soon the two simulations are as different from each other as any other two started independently.

The second data set is the set of Sea Surface Temperatures (SST) used to force the simulations. The data have been compiled in monthly means on the same grid as the atmospheric data. The SST force a special signal on the geopotential field, leaving an identifiable signature in the atmosphere. The variability of the Z field is therefore composed of variability that is intrinsic to the atmosphere and maybe other components of the Earth system and variability that is induced by the SST variability, that varies more slowly. This mixture creates a very rich and challenging situation for the methods presented here. In the last chapter we will use two data sets obtained from simulations from a climate models. They represent the tropical SST and the east-west wind and they also are monthly means for about 200 years of simulations. This is a case of tightly coupled fields that show how the generalized regression methods can really identify covarying fields.

The Empirical Orthogonal Functions are introduced in Chap. 4, after two introductory chapters on basic algebra and basic statistics that are needed to refresh elementary notions and fix the vocabulary. Extensions to the EOF concept are presented in Chap. 5. After having introduced the concept of rotation, we discuss then how EOF can be extended to analyze travelling signals, introducing complex and extended EOF and finally we discuss combined EOF that an effective introduction to the advanced methods of analyzing covariations in Chaps. 6 and 7. A final generalization using linear regression methods is then discussed in Chap. 8.

Chapter 2
Elements of Linear Algebra

2.1 Introduction

This book will use some basic concepts of linear algebra. In this chapter we will briefly recall the essential elements that will be necessary in the rest of the book to understand the various tools devoted to the analysis of variance. Readers that are interested in the more detailed treatment of the subject are directed, for instance, to Golub and Van Loan (1996), Horn and Johnson (1991), Meyer (2000).

2.2 Elementary Vectors

The typical geometric definition introduces a vector as a segment emanating from the origin, with an arrow at the second extreme, indicating a "pointing" direction, or orientation (cf. Fig. 2.1), showing that a vector may be characterized by two properties: length and direction. Although this definition is usually employed on the plane, the same characterization can be used in higher dimension, that is on hyper-planes (e.g. space). While magnitude and direction, for instance, would be sufficient to uniquely identify a vector on the plane, this is not so in higher dimensions. In high dimensions, it is thus more appropriate to characterize vectors by means of their "components". In Cartesian coordinates, these are the orthogonal projections of the vector on each Cartesian axis. Rigorously speaking, a vector is given by an *ordered* n-uple of real or complex numbers, that is, $\mathbf{b} = (b_1, \ldots, b_n)$ is a (row) vector with n components, where each b_i is a real or complex number. Note that the order of the components is important, so that, e.g., the vector $\mathbf{a} = (1, 3)$ is different from the vector $\mathbf{b} = (3, 1)$. The ensemble of all possible vectors is then identified by the ensemble of all possible n-uples of numbers that can be formed with real or complex numbers. A vector with only one component is called a scalar. The vector $\mathbf{0} = (0, \ldots, 0)$ is the zero vector. It is customary to identify the whole of the real and complex numbers with the symbols $\mathbb{R}$ and $\mathbb{C}$, respectively. It thus follows, for instance, that the set of all possible couples is denoted by the symbol $\mathbb{R} \times \mathbb{R} = \mathbb{R}^2$. In general, $\mathbb{R}^n$ is the set of vectors having n components.

A. Navarra and V. Simoncini, *A Guide to Empirical Orthogonal Functions for Climate Data Analysis*, DOI 10.1007/978-90-481-3702-2_2,

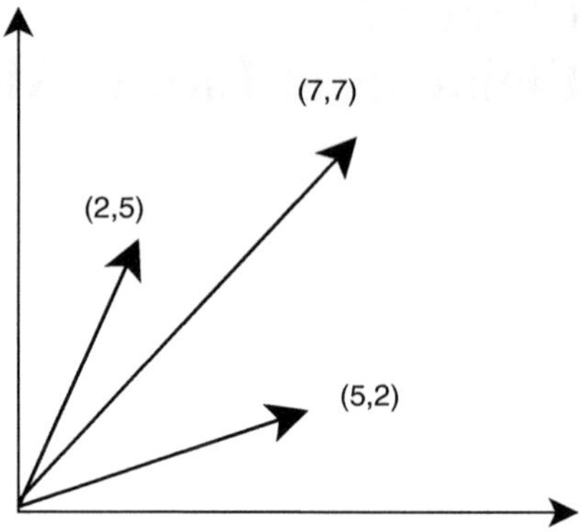

Fig. 2.1 Elementary vectors on the plane

The addition between two vectors *with the same number of components*, is defined as the vector whose components are the sum of the corresponding vector components. If $\mathbf{a} = (a_1, a_2, \ldots, a_n)$ and $\mathbf{b} = (b_1, b_2, \ldots, b_n)$, then $\mathbf{c} := \mathbf{a} + \mathbf{b}$ as $\mathbf{c} = (a_1 + b_1, a_2 + b_2, \ldots, a_n + b_n)$. Note that $\mathbf{c}$ is a vector of n components. By again acting at the component level, we can stretch a vector by multiplying it with a scalar k: we can define $\mathbf{c} = \mathbf{a}k$ where $\mathbf{c} = (ka_1, ka_2, ..., ka_n)$, meaning that each component is multiplied by the factor k. These are the basic operations that allow us to generate a key space for our analysis. In particular, $\mathbb{R}^n$ is closed with respect to the sum and with respect to multiplication by a real scalar, which means that the result of these operations is still an element of $\mathbb{R}^n$. A real vector space is a set that is closed with respect to the addition and multiplication by a real scalar. Therefore, $\mathbb{R}^n$ is a real vector space. There are more complex instances of vector spaces, but for the moment we will content ourselves with this fundamental example. An immediate generalization is given by the definition of a real vector subspace, which is a subset of a real vector space.

2.3 Scalar Product

We next introduce an operation between two vectors that provides the main tool for a geometric interpretation of vector spaces. Given two real vectors $\mathbf{a} = (a_1, a_2, \ldots, a_n)$ and $\mathbf{b} = (b_1, b_2, \ldots, b_n)$, we define the scalar product (or inner product) the operation $\langle \mathbf{a}, \mathbf{b} \rangle = a_1 b_1 + a_2 b_2 + \cdots + a_n b_n$. Note that the operation is between vectors, whereas the result is a real scalar. We remark that if $\mathbf{a}$ and $\mathbf{b}$ were complex vectors, that is vectors with complex components, then a natural inner product would be defined in a different way, and in general, the result would be a complex number (see end of section). The real inner product inherits many useful properties from the product and sum of real numbers. In particular, for any vector $\mathbf{a}, \mathbf{b}, \mathbf{c}$ with n real components and for any real scalar k, it holds

1. Commutative property: $\langle \mathbf{a}, \mathbf{b} \rangle = \langle \mathbf{b}, \mathbf{a} \rangle$
2. Distributive property: $\langle (\mathbf{a} + \mathbf{c}), \mathbf{b} \rangle = \langle \mathbf{a}, \mathbf{b} \rangle + \langle \mathbf{a}, \mathbf{c} \rangle$
3. Multiplication by scalar: $\langle (k\mathbf{a}), \mathbf{b} \rangle = k \langle \mathbf{a}, \mathbf{b} \rangle = \langle \mathbf{a}, (k\mathbf{b}) \rangle$

The scalar product between a vector and itself is of great interest, that is

$$\langle \mathbf{a}, \mathbf{a} \rangle = a_1^2 + a_2^2 + \cdots + a_n^2.$$

Note that $\langle \mathbf{a}, \mathbf{a} \rangle$ is always non-negative, since it is the sum of non-negative numbers. For $n = 2$ it is easily seen from Fig. 3.1 that this is the square of the length of a vector. More generally, we define the Euclidean norm (or simply norm) as

$$||\mathbf{a}|| = \sqrt{\langle \mathbf{a}, \mathbf{a} \rangle}.$$

A versor is a vector of unit norm. Given a non-zero vector $\mathbf{x}$, it is always possible to determine a versor $\mathbf{x}'$ by dividing $\mathbf{x}$ by its norm, that is $\mathbf{x}' = \mathbf{x}/||\mathbf{x}||$. This is a standard form of normalization, ensuring that the resulting vector has norm one. Other normalizations may require to satisfy different criteria, such as, e.g., the first component equal to unity. If not explicitly mentioned, we shall always refer to normalization to obtain unit norm vectors. Given a norm, in our case the Euclidean norm, the distance associated with this norm is

$$d(\mathbf{a}, \mathbf{b}) = ||\mathbf{a} - \mathbf{b}|| = \sqrt{(a_1 - b_1)^2 + (a_2 - b_2)^2 + \cdots + (a_n - b_n)^2}.$$

Scalar products and the induced distance can be defined in several ways; here we are showing only what we shall mostly use in this text. Any function can be used as a norm as long as it satisfies three basic relations: (i) Non-negativity: $\|\mathbf{a}\| \geq 0$ and $\|\mathbf{a}\| = 0$ if and only if $\mathbf{a} = \mathbf{0}$; (ii) Commutative property: $d(\mathbf{a}, \mathbf{b}) = d(\mathbf{b}, \mathbf{a})$; (iii) Triangular inequality: $d(\mathbf{a}, \mathbf{b}) \leq d(\mathbf{a}, \mathbf{c}) + d(\mathbf{c}, \mathbf{b})$. Using norms we can distinguish between close vectors and far away vectors, in other words we can introduce a topology in the given vector space. In particular, property (i) above ensures that identical vectors ($\mathbf{a} = \mathbf{b}$) have a zero distance. As an example of the new possibility offered by vector spaces, we can go back to Fig. 2.1 and consider the angles α and β that the vectors $\mathbf{a}$ and $\mathbf{b}$ in $\mathbb{R}^2$ make with the reference axes. These angles can be easily expressed in terms of the components of the vectors,

$$\cos\beta = \frac{b_1}{\sqrt{b_1^2 + b_2^2}} \qquad \sin\beta = \frac{b_2}{\sqrt{b_1^2 + b_2^2}},$$

$$\cos\alpha = \frac{a_1}{\sqrt{a_1^2 + a_2^2}} \qquad \sin\alpha = \frac{a_2}{\sqrt{a_1^2 + a_2^2}},$$

and also the angle between the two vectors, $\cos(\beta - \alpha)$,

$$\begin{aligned}\cos(\beta - \alpha) &= \cos\beta\cos\alpha + \sin\beta\sin\alpha \\ &= \frac{b_1}{\sqrt{b_1^2 + b_2^2}} \frac{a_1}{\sqrt{a_1^2 + a_2^2}} + \frac{b_2}{\sqrt{b_1^2 + b_2^2}} \frac{a_2}{\sqrt{a_1^2 + a_2^2}},\end{aligned} \tag{2.1}$$

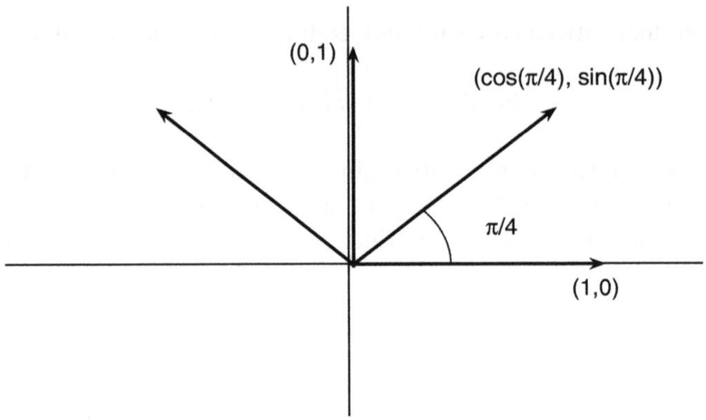

Fig. 2.2 Angles between vectors

or, equivalently,

$$\cos(\beta - \alpha) = \frac{\langle \mathbf{a}, \mathbf{b} \rangle}{||\mathbf{a}||\,||\mathbf{b}||}. \tag{2.2}$$

We just proved that this relation holds in $\mathbb{R}^2$. In higher dimension, the cosine of the angle of two vectors is defined as the ratio between their inner product and their norms, which is nothing but (2.2).

We explicitly observe that the scalar product of two versors gives directly the cosine of the angle between them. By means of this new notion of angle, relative direction of vectors can now be expressed in terms of the scalar product. We say that two vectors are orthogonal if their inner product is zero. Formula (2.2) provides a geometric justification for this definition, which can be explicitly derived in $\mathbb{R}^2$, where orthogonality means that the angles between the two vectors is $\pi/2$ radians (90°); cf. Fig. 2.2. If in addition the two vectors are in fact versors, they are said to be orthonormal.

We can also introduce another geometric interpretation of scalar products that follows from (2.2). The scalar product is also the projection of the vector **a** on **b**: from Fig. 2.3 and from the definition of the cosine the projection of **a** onto the direction of **b** is $Proja = ||\mathbf{a}|| \cos\phi$. Analogously, the projection of **b** onto the direction of **a** is $Projb = ||\mathbf{b}|| \cos\phi$. For normalized vectors the norm disappears and the scalar product gives directly the projections, that are obviously the same in both cases (bottom panel in Fig. 2.3). We close this section with the definition of inner product in the case of complex vectors. Let $\mathbf{x}, \mathbf{y}$ be vectors in $\mathbb{C}^n$. Then $\langle \mathbf{x}, \mathbf{y} \rangle = \bar{x}_1 y_1 + \bar{x}_2 y_2 + \cdots + \bar{x}_n y_n$, where $\bar{x} = a - ib$ denotes the complex conjugate of $x = a + ib$, $i = \sqrt{-1}$. With this definition, the norm of a complex vector is defined as $||\mathbf{x}||^2 = \langle \mathbf{x}, \mathbf{x} \rangle = |x_1|^2 + \cdots + |x_n|^2$.

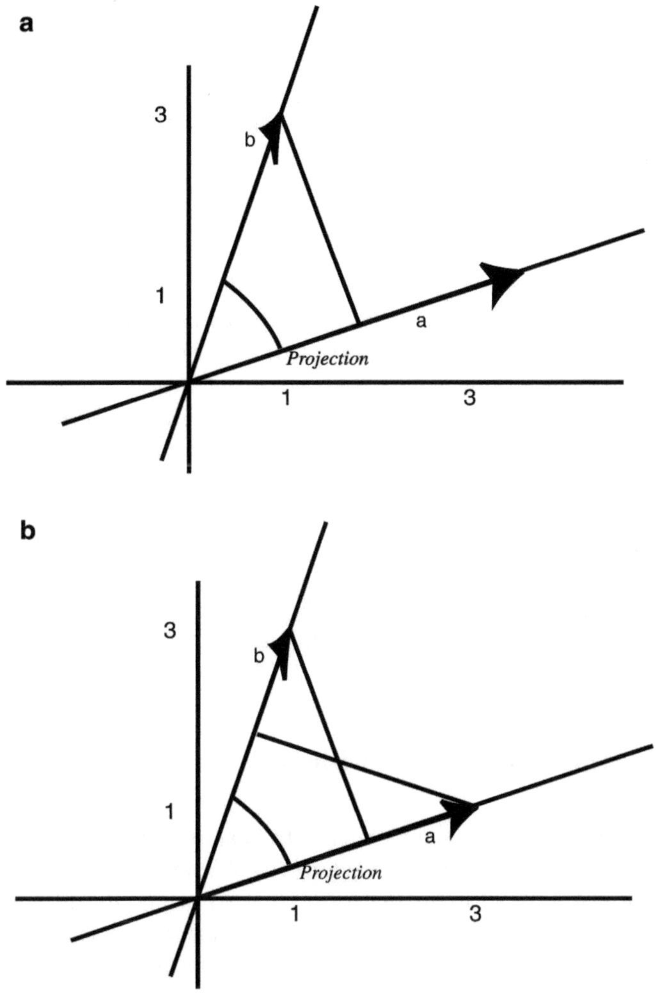

Fig. 2.3 Angles between vectors

Exercises and Problems

1. Given the two vectors $\mathbf{a} = (1, -2, 0)$, $\mathbf{b} = (-3, -1, 4)$, compute $\mathbf{a} + \mathbf{b}$, $\mathbf{a} - \mathbf{b}$, $\mathbf{a} + 2\mathbf{b}$ and $\langle \mathbf{a}, \mathbf{b} \rangle$.
   We have $\mathbf{a}+\mathbf{b} = (1-3, -2-1, 0+4) = (-2, -3, 4)$, $\mathbf{a}-\mathbf{b} = (1+3, -2+1, 0-4) = (4, -1, -4)$ *and* $\mathbf{a}+2\mathbf{b} = (1+2(-3), -2+2(-1), 0+2(4)) = (-4, -4, 8)$. *Finally, we have* $\langle \mathbf{a}, \mathbf{b} \rangle = 1(-3) + (-2)(-1) + 0(4) =$ *-3+2+0=-1.*
2. Given the two complex vectors $\mathbf{x} = (1 + i, -2 + 3i)$, $\mathbf{y} = (-5 + i, 4i)$, compute $\mathbf{x} + \mathbf{y}$ and $\langle \mathbf{x}, \mathbf{y} \rangle$.

We have $\mathbf{x}+\mathbf{y} = (1-5+(1+1)i, -2+(3+4)i) = (-4+2i, -2+7i)$. *Moreover, using the definition of the product between complex numbers,* $\langle \mathbf{x}, \mathbf{y} \rangle = (1-i)(-5+i)+(-2-3i)(4i) = 8-2i$.

3. Given the vectors $\mathbf{a} = (1,-2,1)$ and $\mathbf{b} = (0,2,-3)$, compute $||\mathbf{a}||$, $||\mathbf{b}||$ and $||\mathbf{a}-\mathbf{b}||$. Moreover, normalize $\mathbf{a}$ so as to have unit norm.
We have $||\mathbf{a}|| = (1^2+(-2)^2+1^2)^{1/2} = 6^{1/2}$ *and* $||\mathbf{b}|| = (0^2+2^2+(-3)^2)^{1/2} = 13^{1/2}$. *Moreover,* $||\mathbf{a}-\mathbf{b}|| = ((1-0)^2+(-2-2)^2+(1-3)^2)^{1/2} = 21^{1/2}$. *Finally,* $\mathbf{a}_i = \mathbf{a}/||\mathbf{a}|| = (1/6)^{1/2}(1,-2,1)$.
4. Check whether the following operations or results are admissible: (i) $\mathbf{x}+\mathbf{y}$, with $\mathbf{x} = (1,-1)$, $\mathbf{y} = (1,2,0)$; (ii) $\langle \mathbf{x}, \mathbf{y} \rangle$ with $\mathbf{x}$ and $\mathbf{y}$ as in (i); (iii) $||\mathbf{a}|| = -1$; (iv) $\langle \mathbf{a}, \mathbf{b} \rangle = -\langle \mathbf{b}, \mathbf{a} \rangle$, with $\mathbf{a}, \mathbf{b}$ real vectors of equal dimension; (v) $d(\mathbf{c}, \mathbf{d}) = -1.5$.
None of the statement above is correct. (i) $\mathbf{x}$ *and* $\mathbf{y}$ *have a different number of components hence the two vectors cannot be added. (ii) Same as in (i). (iii) The norm of any vector is non-negative, therefore it cannot be equal to -1. (iv) The inner product of real vectors is commutative, therefore* $\langle \mathbf{a}, \mathbf{b} \rangle = \langle \mathbf{b}, \mathbf{a} \rangle$. *(v) Same as in (iii).*
5. Compute the cosine of the angle between the vectors $\mathbf{a} = (-1,2)$ and $\mathbf{b} = (-3,0)$.
We first compute $\langle \mathbf{a}, \mathbf{b} \rangle = -1(-3)+2(0) = 3$, $||\mathbf{a}|| = \sqrt{5}$ *and* $||\mathbf{b}|| = 3$, *from which we obtain* $\cos\phi = \langle \mathbf{a}, \mathbf{b} \rangle / (||\mathbf{a}|| \, ||\mathbf{b}||) = \frac{1}{\sqrt{5}}$.

2.4 Linear Independence and Basis

Some vectors can be combined and stretched by scalars, hence they can be obtained one from the other. For instance, the vector $(4,4,4)$ can be obtained as $(1,1,1)\cdot 4$ in such a way that all vectors of the form (k,k,k) are really different stretched versions of the same vector $(1,1,1)$. Vectors that cannot be reached with a simple stretching can be obtained with a combination, for instance the vector $(5,2)$ can be written as $2\cdot(1,1)+3\cdot(1,0)$. With this simple example we see that we can choose some particularly convenient vectors to represent all other vectors in the given space. Given r nonzero vectors $\mathbf{x}_1, \mathbf{x}_2, \ldots, \mathbf{x}_r$, we say that a vector $\mathbf{x}$ is a linear combination of these r vectors if there exist r scalars $\alpha_1, \ldots, \alpha_r$, not all equal to zero, such that

$$\mathbf{x} = \alpha_1\mathbf{x}_1 + \alpha_2\mathbf{x}_2 + \cdots + \alpha_r\mathbf{x}_r.$$

This definition is used to distinguish between linearly dependent and independent vectors. In particular, $\mathbf{x}_1, \mathbf{x}_2, \ldots, \mathbf{x}_r$ are said to be linearly dependent if there exist r scalars, not all equal to zero, such that $\alpha_1\mathbf{x}_1 + \alpha_2\mathbf{x}_2 + \cdots + \alpha_r\mathbf{x}_r = \mathbf{0}$. In other words, they are linearly dependent if one of the vectors can be expressed as a linear combination of the other vectors. We are thus ready to define linearly independent vectors, and the associated concept of a basis of a vector space. We say that r vectors $\mathbf{x}_1, \mathbf{x}_2, \ldots, \mathbf{x}_r$ are linearly independent if the only linear combination that gives the zero vector is obtained by setting all scalars equal to zero, that is if the relation $\alpha_1 \mathbf{x}_1 + \alpha_2 \mathbf{x}_2 + \cdots + \alpha_r \mathbf{x}_r = \mathbf{0}$ implies $\alpha_1 = \alpha_2 = \cdots = \alpha_r = 0$. The maximum number of linearly independent vectors is called the dimensionality of the vector space, maybe is not surprising that for $\mathbb{R}^n$ this number turns out to be n. Given n linearly independent vectors in $\mathbb{R}^n$, any other vector can be obtained as a linear combination of these n vectors. For this reason, n linearly independent vectors of $\mathbb{R}^n$, are called a basis of $\mathbb{R}^n$. Clearly, a basis is not uniquely determined, since any group of n linearly independent vectors represents a basis. However, choices that are particularly convenient are given by sets of normalized and mutually orthogonal and thus independent vectors, namely, we select an "orthonormal basis". In the case of $\mathbb{R}^2$, orthonormal bases are for instance $(1, 0)$, $(0, 1)$, and also $(1/\sqrt{2}, 1/\sqrt{2})$, $(1/\sqrt{2}, -1/\sqrt{2})$. In fact, the latter can be obtained from the first one with a rotation, as it is shown in Fig. 2.2.

The orthonormal basis $\mathbf{e}_1, \mathbf{e}_2, \ldots, \mathbf{e}_n$, where $\mathbf{e}_k = (0, 0, \ldots, 1, 0, \ldots, 0)$, that is, all components are zero except the unit kth component, is called the canonical basis of $\mathbb{R}^n$. Note that it is very simple to obtain the coefficients in the linear combination of a vector of $\mathbb{R}^n$ in terms of the canonical basis: these coefficients are simply the components of the vector (see Exercise 3 below). The choice of a particular basis is mainly dictated by either computational convenience or by ease of interpretation. Given two vectors $\mathbf{x}, \mathbf{y}$ in $\mathbb{R}^n$, it is always possible to generate a vector from $\mathbf{x}$, that is orthogonal to $\mathbf{y}$. This goes as follows: we first define the vector $\mathbf{y}' = \mathbf{y}/||\mathbf{y}||$ and the scalar $t = \langle \mathbf{y}', \mathbf{x} \rangle$, with which we form $\mathbf{x}' = \mathbf{x} - \mathbf{y}'t$. The computed $\mathbf{x}'$ is thus orthogonal to $\mathbf{y}$. Indeed, using the properties of the inner product, $\langle \mathbf{y}', \mathbf{x}' \rangle = \langle \mathbf{y}', \mathbf{x} - \mathbf{y}'t \rangle = \langle \mathbf{y}', \mathbf{x} \rangle - t\langle \mathbf{y}', \mathbf{y}' \rangle = t - t = 0$.

Determining an orthogonal basis of a given space is a major task. In $\mathbb{R}^2$ this is easy: given any vector $\mathbf{a}=(a_1, a_2)$, the vector $\mathbf{b}=(-a_2, a_1)$ (or $\mathbf{c}=-\mathbf{b} = (a_2, -a_1)$) is orthogonal to $\mathbf{a}$, therefore the vectors $\mathbf{a}, \mathbf{b}$ readily define an orthogonal basis. In $\mathbb{R}^n$ the process is far less trivial. A stable way to proceed is to take n linearly independent vectors $\mathbf{u}_1, \ldots, \mathbf{u}_n$ of $\mathbb{R}^n$, and then orthonormalize them in a sequential manner. More precisely, we first normalize $\mathbf{u}_1$ to get $\mathbf{v}_1$; we take $\mathbf{u}_2$, we orthogonalize it against $\mathbf{v}_1$ and then normalize it to get $\mathbf{v}_2$. We thus continue with $\mathbf{u}_3$, orthogonalize it against $\mathbf{v}_1$ and $\mathbf{v}_2$ and get $\mathbf{v}_3$ after normalization, and so on. This iterative procedure is the famous Gram-Schmidt process.

Exercises and Problems

1. Given the two vectors $\mathbf{a} = (-1, -2)$ and $\mathbf{b} = (-3, -1)$: (i) verify that $\mathbf{a}$ and $\mathbf{b}$ are linearly independent. (ii) Compute a vector orthogonal to $\mathbf{a}$. (iii) If possible, determine a scalar k such that $\mathbf{c} = k\mathbf{a}$ and $\mathbf{a}$ are linearly independent.
   (i) In $\mathbb{R}^2$ vectors are either multiple of each other or they are independent. Since $\mathbf{b}$ *is not a multiple of* $\mathbf{a}$, *we have that* $\mathbf{a}$ *and* $\mathbf{b}$ *are linearly independent. (ii) The vector* $\mathbf{d} = (2, -1)$ *is orthogonal to* $\mathbf{a}$, *indeed* $\langle \mathbf{a}, \mathbf{d} \rangle = (-1)(2) + (-2)(-1) = 0$. *(iii) From the answer to (i), it follows that there is no such* $\mathbf{c}$.
2. Obtain an orthonormal set from the two linearly independent vectors: $\mathbf{a} = (2, 3)$ and $\mathbf{b} = (1, 1)$.

We use the Gram-Schmidt process. First, $\mathbf{a}' = \mathbf{a}/||\mathbf{a}|| = \frac{1}{\sqrt{13}}(2,3)$*. Then, we compute* $\langle \mathbf{a}', \mathbf{b}\rangle = \frac{5}{\sqrt{13}}$*, so that* $\mathbf{b} = \mathbf{b} - \mathbf{a}'\langle \mathbf{a}', \mathbf{b}\rangle = (1,1) - \frac{5}{\sqrt{13}}\mathbf{a}' = (1,1) - \frac{5}{13}(2,3) = \frac{1}{13}(3,-2)$*, from which* $\mathbf{b}' = \mathbf{b}/||\mathbf{b}|| = \frac{1}{2\cdot 13}(3,-2)$*. Hence,* $\mathbf{a}', \mathbf{b}'$ *is the sought after set. Not surprisingly (cf. text),* $\mathbf{b}'$ *is the normalized version of* $(-3,2)$*, easily obtainable directly from* $\mathbf{a}$*.*

3. Given the vector $\mathbf{x} = (3,-2,4)$, determine the coefficients in the canonical basis.
   We simply have $\mathbf{x} = 3\mathbf{e}_1 - 2\mathbf{e}_2 + 4\mathbf{e}_3$.
4. By simple inspection, determine a vector that is orthogonal to each of the following vectors: $\mathbf{a} = (1,0,0,1), \mathbf{b} = (4,3,1,0,0), \mathbf{c} = (0.543, 1.456, 1, 1)$.
   It can be easily verified that any of the vectors $(0,\alpha,\beta,0)$, $(-1,\alpha,\beta,1)$, $(1,\alpha,\beta,-1)$, *with* α,β *scalars, are orthogonal to* $\mathbf{a}$*. Analogously,* $(0,0,0,\alpha,\beta)$ *are orthogonal to* $\mathbf{b}$*, together with* $(-1,1,1,\alpha,\beta)$, $(1,-1,-1,\alpha,\beta)$.
   For $\mathbf{c}$*, simple choices are* $(0,0,-1,1)$ *and* $(0,0,1,-1)$.

2.5 Matrices

A matrix is an $n \times m$ rectangular array of scalars, real or complex numbers, with n rows and m columns. When $m = n$ the matrix is "square" and n is its dimension. In this book, we will use capital bold letters to indicate matrices, whereas roman small case letters in bold are used to denote vectors; Greek letters will commonly denote scalars. The following are examples of matrices of different dimensions,

$$\mathbf{A} = \begin{pmatrix} 0 & -1 & 4 \\ \frac{1}{2} & 2 & 1 \end{pmatrix}, \quad \mathbf{B} = \begin{pmatrix} 0 & 1 \\ i & 0 \end{pmatrix}, \quad \mathbf{C} = \begin{pmatrix} 1 & 0 \\ 0 & 1+2i \\ 0.05 & -1 \\ 1.4+5i & 2 \\ 0 & 3 \end{pmatrix}. \tag{2.3}$$

Matrix $\mathbf{A}$ is 2×3, $\mathbf{B}$ is 2×2 and $\mathbf{C}$ is 5×2. Note that $\mathbf{B}$ and $\mathbf{C}$ have complex entries. The components of a matrix $\mathbf{A}$ are denoted by $a_{i,j}$, where i corresponds to the ith row and j to the jth column, that is at the (i,j) position in the array. In the following we shall use either parentheses or brackets to denote matrices. The $n \times m$ matrix with all zero entries is called the zero matrix. The square matrix with ones at the (i,i) entries, $i = 1, \ldots, n$ and zero elsewhere, is called the identity matrix and is denoted by $\mathbf{I}$. If the order is not clear from the context, we shall use $\mathbf{I}_n$. The position of the scalars within the array is important: matrices with the same elements, but in a different order, are distinct matrices. Of particular interest is the transpose matrix, i.e. the matrix $b\mathbf{A}^T$ obtained by exchanging rows and columns of the matrix $\mathbf{A}$. For instance, for the matrices in (2.3),

$$\mathbf{A}^T = \begin{pmatrix} 0 & \frac{1}{2} \\ -1 & 2 \\ 4 & 1 \end{pmatrix}, \quad \mathbf{B}^T = \begin{pmatrix} 0 & i \\ 1 & 0 \end{pmatrix},$$

$$\mathbf{C}^T = \begin{pmatrix} 1 & 0 & 0.05 & 1.4+5i & 0 \\ 0 & 1+2i & -1 & 2 & 3 \end{pmatrix},$$

are the transpose matrices of the previous example. In the case of matrices with complex entries, we can also define the complex transposition, indicated by the subscript '$*$', obtained by taking the complex conjugate of each element of the transpose, $\mathbf{B}^* := \bar{\mathbf{B}}^T$, so that

$$\mathbf{B}^* = \begin{pmatrix} 0 & -i \\ 1 & 0 \end{pmatrix}, \quad \mathbf{C}^* = \begin{pmatrix} 1 & 0 & 0.05 & 1.4-5i & 0 \\ 0 & 1-2i & -1 & 2 & 3 \end{pmatrix}.$$

Clearly, for real matrices the *Hermitian adjoint* $\mathbf{B}^*$ coincides with the transpose matrix. Transposition and Hermitian adjoint share the reverse order law, i.e. $(\mathbf{AB})^* = \mathbf{B}^*\mathbf{A}^*$ and $(\mathbf{AB})^T = \mathbf{B}^T\mathbf{A}^T$, where $\mathbf{A}$ and $\mathbf{B}$ have conforming dimensions. See later for the definition of matrix-matrix products. Matrices that satisfy $\mathbf{A}^*\mathbf{A} = \mathbf{A}\mathbf{A}^*$ are called *normal*. A real square matrix $\mathbf{X}$ such that $\mathbf{X}^T\mathbf{X} = \mathbf{I}$ and $\mathbf{X}\mathbf{X}^T = \mathbf{I}$ is said to be an *orthogonal matrix*. A square complex matrix $\mathbf{X}$ such that $\mathbf{X}^*\mathbf{X} = \mathbf{I}$ and $\mathbf{X}\mathbf{X}^* = \mathbf{I}$ is said to be unitary.

An $n \times n$ matrix $\mathbf{A}$ is *invertible* if there exists a matrix $\mathbf{B}$ such that $\mathbf{AB} = \mathbf{BA} = \mathbf{I}$. If such a matrix $\mathbf{B}$ exists, it is unique, and it is called the *inverse* of $\mathbf{A}$, and it is denoted by $\mathbf{A}^{-1}$. An invertible matrix is also called nonsingular. Therefore, a singular matrix is a matrix that is not invertible. Recalling the definition of orthogonal matrices, we can immediately see that an orthogonal matrix is always invertible and more precisely, we have that its inverse coincides with its transpose, that is $\mathbf{X}^T = \mathbf{X}^{-1}$ (for a unitary matrix $\mathbf{X}$, it is $\mathbf{X}^* = \mathbf{X}^{-1}$). Matrices with special structures are given specific names. For instance,

$$\mathbf{D} = \begin{pmatrix} 1 & 0 & 0 & 0 \\ 0 & 2 & 0 & 0 \\ 0 & 0 & 5 & 0 \\ 0 & 0 & 0 & 1 \end{pmatrix},$$

$$\mathbf{U} = \begin{pmatrix} 1 & 1+2i & 4+2i & 3 \\ 0 & 2 & 1 & 3 \\ 0 & 0 & 5 & 4i \\ 0 & 0 & 0 & 1 \end{pmatrix}, \quad \mathbf{L} = \begin{pmatrix} 1 & 0 & 0 & 0 \\ 1-2i & 2 & 0 & 0 \\ 4-2i & 1 & 5 & 0 \\ 3 & 3 & 4i & 1 \end{pmatrix}. \tag{2.4}$$

Note also that in this example, $\mathbf{L} = \mathbf{U}^*$. Matrices like $\mathbf{D}$ are called diagonal (zero entries everywhere but on the main "diagonal"), whereas matrices like $\mathbf{U}$ and $\mathbf{L}$ are called upper triangular and lower triangular, respectively, since only the upper (resp. lower) part of the matrix is not identically zero. Note that it is very easy to check whether a diagonal matrix is invertible. Indeed, it can be easily verified that a diagonal matrix with diagonal components the inverses of the original diagonal entries is the sought after inverse. For the example above, we have $\mathbf{D}^{-1} = \text{diag}(1, \frac{1}{2}, \frac{1}{5}, 1)$ (we have used here a short-hand notation, with obvious meaning). Therefore, if all diagonal entries of $\mathbf{D}$ are nonzero, $\mathbf{D}$ is invertible, and vice versa. Similar considerations can be applied for (upper or lower) triangular matrices, which are nonsingular if and only if their diagonal elements are nonzero. Explicitly determining the inverse of less structured matrices is a much more difficult task. Fortunately, in most applications this problem can be circumvented.

If we look closely at the definition of a matrix we can see that there are several analogies to the definition of vectors we have used in the preceding sections. In fact we can think of each column as a vector, for instance the first column of the matrix $\mathbf{C}$ is the vector $(1, 0, 0.05, 1.4 + 5i, 0)$ a vector of the four-dimensional vector space $\mathbb{C}^4$. More generally, any row or column of $\mathbf{C}$ can be viewed as a single vector. In the following we will need both kinds of vectors, but we will follow the convention that we will use the name "vector" for the column orientation, i.e. matrices with dimension $n \times 1$. Since a vector is just a skinny matrix, we can go from a column to a row vector via a transposition:

$$\mathbf{u} = \begin{pmatrix} 1 \\ 2 \\ 3 \end{pmatrix}, \qquad \mathbf{u}^T = (1\ 2\ 3).$$

From now on, the use of row vectors will be explicitly indicated by means of the transposition operation.

It can be shown that matrices are representations of linear transformations connecting vector spaces. In other words, an $m \times n$ matrix $\mathbf{M}$ is an application that maps a vector $\mathbf{u}$ of an n-dimensional vector space $\mathcal{U}$ onto an element $\mathbf{v}$ of an m-dimensional vector space $\mathcal{V}$, that is

$$\mathbf{v} = \mathbf{M}\mathbf{u}.$$

The vector space $\mathcal{U}$ is known as the *domain* of $\mathbf{M}$ and the vector space $\mathcal{V}$ is called the *range*. Another important vector space associated with a matrix $\mathbf{M}$ is the null space, i.e. the subset of the domain such that for all $\mathbf{u}$ in this space, it holds $\mathbf{M}\mathbf{u} = \mathbf{0}$.

The product of a matrix $\mathbf{A} = (a_{i,j})$ on a vector is defined as another vector $\mathbf{v} = \mathbf{A}\mathbf{u}$ whose components are obtained by the row-by-column multiplication rule

$$v_i = \sum_{j=1}^{m} a_{i,j} u_j, \quad i = 1, \ldots, n.$$

It can be noticed that the ith component of the resulting vector $\mathbf{v}$ is the scalar product of the ith row of $\mathbf{A}$ with the given vector $\mathbf{u}$. The matrix–vector product rule can be used to define a matrix–matrix multiplication rule by repeatedly applying the rule to each column of the second matrix, such that the element of the product matrix $\mathbf{C} = \mathbf{AB}$ of the $q \times m$ matrix $\mathbf{A}$ with the $m \times p$ matrix $\mathbf{B}$ is given by

$$c_{i,j} = \sum_{k=1}^{m} a_{i,k} b_{k,j}, \quad i = 1, \ldots, q, \quad j = 1, \ldots, p.$$

Note that the resulting matrix $\mathbf{C}$ has dimension $q \times p$. For the product to be correctly defined, the number of columns of the first matrix, in this case $\mathbf{A}$, must be equal to the number of rows of the second matrix, $\mathbf{B}$. Note that this operation is not commutative, that is, in general $\mathbf{AB} \neq \mathbf{BA}$, even if both products are well defined. On the other hand, matrices that do satisfy the commutative property are said to commute with each other. Subsets of matrices that commute with each other have special properties that will appear in the following. A simple class of commuting matrices is given by the diagonal matrices: if $\mathbf{D}_1$ and $\mathbf{D}_2$ are diagonal matrices, then it can be verified that it always holds that $\mathbf{D}_1\mathbf{D}_2 = \mathbf{D}_2\mathbf{D}_1$.

Our last basic fact concerning matrices is related to the generalization to matrices of the vector notion of norm. In particular, we will use the *Frobenius norm*, which is natural generalization to matrices of the Euclidean vector norm. More precisely, given $b\mathbf{A} \in \mathbb{R}^{n \times m}$,

$$\|\mathbf{A}\|_F^2 := \sum_{j=1}^{m} \sum_{i=1}^{n} a_{i,j}^2, \tag{2.5}$$

which can be equivalently written as $\|\mathbf{A}\|_F^2 == \sum_{j=1}^{m} \|\mathbf{a}_j\|^2$, where $\mathbf{a}_j$ is the jth column of $\mathbf{A}$ (a corresponding relation holds for the rows). In particular, it holds that $\|\mathbf{A}\|^2 = \text{trace}(\mathbf{A}^T\mathbf{A})$, where the trace of a matrix is the sum of its diagonal elements. The definition naturally generalizes to complex matrices. It is also interesting that the Frobenius norm is invariant under rotations, that is the norm remains unchanged whenever we multiply an orthonormal matrix by the given matrix. In other words, for any orthonormal matrix $\mathbf{Q}$ it holds that $\|\mathbf{A}\|_F = \|\mathbf{QA}\|_F$.

Exercises and Problems

1. Given the matrices $\mathbf{A} = \begin{pmatrix} -1 & -3 \\ -1 & 2 \end{pmatrix}$ and $\mathbf{B} = \begin{pmatrix} 0 & -1 \\ -3 & 1 \end{pmatrix}$, compute $\mathbf{AB}$.

   We have

$$\boldsymbol{AB} = \begin{pmatrix} -1 & -3 \\ -1 & 2 \end{pmatrix} \begin{pmatrix} 0 & -1 \\ -3 & 1 \end{pmatrix}$$

$$= \begin{pmatrix} -1(0) - 3(-3) & -1(1) - 3(1) \\ -1(0) + 2(-3) & -1(-1) + 2(1) \end{pmatrix} = \begin{pmatrix} 9 & -4 \\ -6 & 3 \end{pmatrix}.$$

2. Given the matrix $\mathbf{A}$ above, compute $\mathbf{A}^T$.
   We have $\mathbf{A}^T = \begin{pmatrix} -1 & -1 \\ -3 & 2 \end{pmatrix}$.
3. Given the matrix $\mathbf{A}$ above, and the row vector $\mathbf{x}^T = (-1, 5)$, compute $\mathbf{x}^T\mathbf{A}$.
   We have $\mathbf{x}^T A = (-1,5)\begin{pmatrix} -1 & -1 \\ -3 & 2 \end{pmatrix} = ((-1)(-1) + 5(-3), (-1)(-1) + 5(2)) = (-14, 11)$.
4. Compute $\mathbf{x}\mathbf{y}^T$ with $\mathbf{x}^T = (1, -3)$ and $\mathbf{y}^T = (2, -2)$.
   The result of this computation is the 2×2 *matrix given by*

$$\mathbf{x}\mathbf{y}^T = \begin{pmatrix} 1 \\ -3 \end{pmatrix}(2, -2) = \begin{pmatrix} 2 & -2 \\ -6 & 6 \end{pmatrix}.$$

   (In the computation, the vectors $\mathbf{x}, \mathbf{y}^T$ *are viewed as* 2×1 *and* 1×2 *matrices, respectively)*

2.6 Rank, Singularity and Inverses

The maximum number of columns or rows that are linearly independent in a matrix $\mathbf{A}$ is called rank, denoted in the following by rank($\mathbf{A}$). For a given $m \times n$ matrix $\mathbf{A}$, clearly rank($\mathbf{A}$) $\leq \min\{n, m\}$. The rank can be used very efficiently to characterize the existence of the solution of a linear system of equations. In matrix terms, a linear system can be written as

$$\mathbf{A}\mathbf{x} = \mathbf{b}, \tag{2.6}$$

where $\mathbf{x}$ represents the vector of the unknown variables, the entries of $\mathbf{A}$ the system's coefficients, and the components of $\mathbf{b}$ are the given right-hand sides of each equation. The system in (2.6) can either have no solution, one solution or infinite solutions. Let us write $\mathbf{A} = (\mathbf{a}_1, \mathbf{a}_2, \ldots, \mathbf{a}_n)$, where $\mathbf{a}_i$, $i = 1, \ldots, n$ are the columns of $\mathbf{A}$. By reading (2.6) backwards, we look for $\mathbf{x} = (x_1, \ldots, x_n)^T$ such that $\mathbf{b} = \mathbf{A}\mathbf{x}$, that is, we seek the coefficients $x_1, \ldots, x_n$, such that $\mathbf{b} = \mathbf{a}_1 x_1 + \cdots + \mathbf{a}_n x_n$. In other words, the solution vector $\mathbf{x}$ yields the coefficients that allow us to write $\mathbf{b}$ as a linear combination of the columns of $\mathbf{A}$. At least one solution exists if rank($\mathbf{A}$) = rank(($\mathbf{A}, \mathbf{b}$)), where ($\mathbf{A}, \mathbf{b}$) is the matrix obtained by adding the vector $\mathbf{b}$ as a column besides $\mathbf{A}$. This corresponds to saying that the $n+1$ vectors $\{\mathbf{a}_1, \mathbf{a}_2, \ldots, \mathbf{a}_n, \mathbf{b}\}$ are linearly dependent. The condition on the rank also shows that the existence of solutions to the system is related to the rank of the coefficient matrix $\mathbf{A}$. For square matrices, using the definition of inverse, $\mathbf{A}\mathbf{x} = \mathbf{b}$ is equivalent to $\mathbf{A}^{-1}\mathbf{A}\mathbf{x} = \mathbf{A}^{-1}\mathbf{b}$, that is $\mathbf{x} = \mathbf{A}^{-1}\mathbf{b}$. Hence, assuming that $\mathbf{b}$ is a nonzero vector, a unique solution $\mathbf{x}$ exists if and only if $\mathbf{A}$ is nonsingular. A crucial result of linear algebra is the following: an $n \times n$ (square) matrix $\mathbf{A}$ is invertible if and only if $\text{rank}(\mathbf{A}) = n$. In particular, this result implies that if $\mathbf{A}$ is singular, the columns of $\mathbf{A}$ are linearly dependent ($\text{rank}(\mathbf{A}) < n$), or equivalently, there exists a vector $\mathbf{x}$, not identically zero, such that $\mathbf{Ax} = \mathbf{0}$. We have thus found that a singular matrix is characterized by a non-empty null space (cf. Sect. 3.5).

2.7 Decomposition of Matrices: Eigenvalues and Eigenvectors

A complex scalar λ and a nonzero complex vector $\mathbf{x}$ are said to be an eigenvalue and an eigenvector of a square matrix $\mathbf{A}$, respectively, if they satisfy

$$\mathbf{Ax} = \lambda\mathbf{x}. \tag{2.7}$$

A vector satisfying (2.7) has the special property that multiplication by $\mathbf{A}$ does not change its direction, but only its length. In the case of Hermitian $\mathbf{A}$ (i.e. $\mathbf{A} = \mathbf{A}^*$), it can be shown that such vectors arise in the problem of maximizing $\langle \mathbf{x}, \mathbf{Ax} \rangle$, over all vectors $\mathbf{x}$ such that $||\mathbf{x}|| = 1$. It is then found that the solution must satisfy the equation $\mathbf{Ax} = \lambda\mathbf{x}$, where λ is a scalar. The pair $(\lambda, \mathbf{x})$ is called an eigenpair of $\mathbf{A}$. The set of all eigenvalues of $\mathbf{A}$ is called the *spectrum* of $\mathbf{A}$. It is important to notice that eigenvectors are not uniquely determined. For instance, if $\mathbf{x}$ is an eigenvector associated with λ, then $\alpha\mathbf{x}$ with $\alpha \neq 0$ is also an eigenvector associated with λ. Finally, we observe that if $\mathbf{A}$ is singular, then there exists a vector $\mathbf{x}$ such that $\mathbf{Ax} = 0 = 0\mathbf{x}$, that is, $\lambda = 0$ is an eigenvalue of $\mathbf{A}$ and $\mathbf{x}$ is the corresponding eigenvector.

A fundamental result is that each square matrix $\mathbf{A}$ of dimension n has exactly n complex eigenvalues, not necessarily all distinct. In case of multiple copies of the same eigenvalue, such a number of copies is called the multiplicity of that eigenvalue.[1] On the one hand, there can be *at most n* linearly independent eigenvectors. If an eigenvalue has multiplicity m larger than one, then there may be at most m linearly independent eigenvectors associated with that eigenvalue. On the other hand, eigenvectors corresponding to different eigenvalues are always linearly independent. Therefore, for a general matrix $\mathbf{A}$, the only case when there may not be a full set of independent eigenvectors is when there are multiple eigenvalues.

The case of Hermitian matrices is particularly fortunate, since in this case, there always exists a set of n linearly independent, and even mutually orthonormal, eigenvectors, regardless of the eigenvalue multiplicity. For a general square matrix $\mathbf{A}$, if there exist n linearly independent eigenvectors, $\mathbf{A}$ can be written as

$$\mathbf{A} = \mathbf{X}\mathbf{\Lambda}\mathbf{X}^{-1}, \tag{2.8}$$

[1] To be more precise, this number is the *algebraic* multiplicity of the eigenvalue.

where $\mathbf{\Lambda}$ is a diagonal matrix having the eigenvalues of $\mathbf{A}$ as diagonal entries, while $\mathbf{X} = [\mathbf{x}_1, \mathbf{x}_2, \ldots, \mathbf{x}_n]$ is a matrix formed by normalized eigenvectors. The inverse of $\mathbf{X}$ exists in this case because we are assuming that the eigenvectors are linearly independent, namely $\mathbf{X}$ has rank n. If a form as in (2.8) can be written, we say that $\mathbf{A}$ is diagonalizable. In the important case of Hermitian matrices, thanks to the orthogonality of the eigenvectors, we can write

$$\mathbf{A} = \mathbf{X}\mathbf{\Lambda}\mathbf{X}^*$$

where $\mathbf{X}$ is the matrix of the eigenvectors, normalized so as to have unit norm. Therefore, for Hermitian matrices, no inversion is required, as $\mathbf{X}^* = \mathbf{X}^{-1}$. If $\mathbf{A}$ is real and symmetric, then the eigenpairs are real.

It can be shown that the eigenvalues can be found by solving the following scalar equation as a function of λ,

$$\det(\mathbf{A} - \lambda\mathbf{I}) = 0, \tag{2.9}$$

whose left-hand side is a polynomial (the *characteristic polynomial*) of degree n in λ. Afterwards, the eigenvectors are obtained by solving the singular system

$$(\mathbf{A} - \lambda_i\mathbf{I})\mathbf{x}_i = 0, \quad i = 1, \ldots, k,$$

where the index i runs over all k distinct eigenvalues found from solving (2.9). From the theory of polynomials, it follows that if $\mathbf{A}$ is real, then its eigenvalues are real or, they appear as complex conjugates, that is, if λ is a complex eigenvalue of $\mathbf{A}$, then $\bar{\lambda}$ is also an eigenvalue of $\mathbf{A}$. Eigenvectors corresponding to real eigenvalues of a real matrix $\mathbf{A}$, can be taken to be real. Finally, Hermitian matrices have only real eigenvalues.

Nondiagonalizable matrices cannot be written in the form (2.8) with $\mathbf{\Lambda}$ diagonal. In particular, a nondiagonalizable matrix of dimension n does not have n linearly independent eigenvectors. This situation may only occur in the presence of multiple eigenvalues (see Exercises 4 and 5 at the end of this chapter).

The transformation indicated by (2.8) is an example of a class of transformations known as similarity transformations. Two matrices $\mathbf{A}$ and $\mathbf{B}$ are said to be similar if they can be obtained from each other by a similarity transformation via a nonsingular matrix $\mathbf{S}$, that is

$$\mathbf{A} = \mathbf{S}\mathbf{B}\mathbf{S}^{-1}. \tag{2.10}$$

Similar matrices share important properties, for instance, they have the same set of eigenvalues. The similarity transformation is equivalent to a change of basis in the representation of the matrix, in fact it can be shown that the transformation (2.10) is equivalent to changing the basis of the column vectors of the matrix $\mathbf{B}$, resulting in different coordinates.

2.8 The Singular Value Decomposition

Square as well as rectangular matrices can always be *diagonalized* if we allow the usage of two transformation matrices instead of one. Any $m \times n$ matrix $\mathbf{A}$ with $m \geq n$, can be decomposed as:

$$\mathbf{A} = \mathbf{U} \begin{pmatrix} \boldsymbol{\Sigma} \\ 0 \end{pmatrix} \mathbf{V}^*, \tag{2.11}$$

where $\mathbf{U} = [\mathbf{u}_1, \mathbf{u}_2, \ldots, \mathbf{u}_m]$ and $\mathbf{V} = [\mathbf{v}_1, \mathbf{v}_2, \ldots, \mathbf{v}_n]$ are square, unitary matrices of dimension m and n, respectively. Matrix $\boldsymbol{\Sigma}$ is diagonal and real, $\boldsymbol{\Sigma} = \mathrm{diag}(\sigma_1, \sigma_2, \ldots, \sigma_n)$, with $\sigma_{i+1} \leq \sigma_i$, $i = 1, \ldots, n-1$, and $\sigma_i \geq 0$, $i = 1, \ldots, n$. A completely analogous decomposition holds for $n \geq m$. The decomposition in (2.11) is called *singular value decomposition* (SVD); the columns of $\mathbf{U}$ and $\mathbf{V}$ are left and right singular vectors, respectively; the real numbers $\sigma_1, \sigma_2, \ldots, \sigma_n$ are called singular values. The following relations can be derived,

$$\mathbf{A}\mathbf{A}^* = \mathbf{U}\boldsymbol{\Sigma}^2\mathbf{U}^*, \quad \mathbf{A}^*\mathbf{A} = \mathbf{V}\boldsymbol{\Sigma}^2\mathbf{V}^*,$$

indicating that the columns of the matrix $\mathbf{U}$ are the eigenvectors of the matrix $\mathbf{A}\mathbf{A}^*$, while the columns of $\mathbf{V}$ are the eigenvectors of the transpose matrix $\mathbf{A}^*\mathbf{A}$. Using the orthogonality of $\mathbf{U}$ and $\mathbf{V}$ in (2.11), we can write

$$\mathbf{A}\mathbf{V} = \mathbf{U} \begin{pmatrix} \boldsymbol{\Sigma} \\ 0 \end{pmatrix}, \quad \mathbf{A}^*\mathbf{U} = \mathbf{V}(\boldsymbol{\Sigma}, 0).$$

If $\mathbf{A}$ is real, then all matrices have real entries. A series of very important results links the SVD with the determination of the rank of matrices. It can be shown that the rank, i.e. the number of linearly independent columns or rows in a matrix, is given by the number of non-zero singular values. The problem of finding the rank of a matrix can therefore be reduced to the problem of finding the number of nonzero singular values. Full rank square matrices of dimension n, have therefore exactly n strictly positive singular values. Comparing (2.8) with (2.11) we can see that the singular values decomposition extends the diagonalization property of the eigenvalues to more general matrices, including rectangular ones. The eigenvalue decomposition looks for a similarity transformation to a diagonal form, whereas in the singular value decomposition, we look for two, in general different, unitary transformations to a diagonal form.

We next briefly discuss the tight connection between the SVD and certain matrix norms that are *induced* by a vector norm. Let $\mathbf{A}$ be an $m \times n$ matrix. Using the Euclidean norm we can define

$$\|\mathbf{A}\|_2 = \max_{\mathbf{x} \neq \mathbf{0}, \mathbf{x} \in \mathbb{C}^n} \frac{\|\mathbf{A}\mathbf{x}\|_2}{\|\mathbf{x}\|_2}.$$

It can be shown that the vector $\mathbf{x}$ that achieves this maximum is the first right singular vector, $\mathbf{v}_1$, so that $\|\mathbf{A}\|_2 = \|\mathbf{A}\mathbf{v}_1\|_2 = \sigma_1$. The SVD allows us to also determine the matrix of low rank that is closest to the original matrix $\mathbf{A}$ in the 2-norm. More precisely, let

$$\mathbf{A}_k = (\mathbf{u}_1, \ldots, \mathbf{u}_k) \begin{pmatrix} \sigma_1 & \cdots & 0 \\ 0 & \ddots & 0 \\ 0 & 0 & \sigma_k \end{pmatrix} (\mathbf{v}_1, \ldots, \mathbf{v}_k)^*$$

be the matrix formed by the first k singular triplets. In other words, $\mathbf{A}_k$ is the matrix obtained by a *truncated* SVD of rank k. Then it holds

$$\min_{\mathbf{B} \in \mathbb{C}^{m \times n}, \mathrm{rank}(\mathbf{B}) = k} \|\mathbf{A} - \mathbf{B}\|_2 = \|\mathbf{A} - \mathbf{A}_k\|_2 = \sigma_{k+1}.$$

The relation above says that $\mathbf{A}_k$ is the rank-k matrix that is closest to $\mathbf{A}$ when using the 2-norm. Moreover, it provides an explicit value for the error of such approximation, which is given by the first neglected singular value, σ_{k+1}.

The SVD can also be employed for computing the Frobenius norm of a matrix; see (2.5). Indeed, it holds that

$$\|\mathbf{A}\|_F^2 = \sum_{j=1}^{\min\{n,m\}} \sigma_j^2,$$

where σ_j's are the singular values of the $n \times m$ matrix $\mathbf{A}$.

The singular value decomposition provides a formidable tool to replace the inverse of a singular or rectangular matrix. Assume that an $m \times n$ matrix $\mathbf{A}$ with $m \geq n$ is decomposed as in (2.11), where $\boldsymbol{\Sigma}$ is nonsingular. Then the *Penrose* pseudo-inverse of $\mathbf{A}$ (cf., e.g., Golub and Van Loan 1996) is defined as[2]

$$\mathbf{A}^- := \mathbf{V} \begin{pmatrix} \boldsymbol{\Sigma}^{-1} & 0 \end{pmatrix} \mathbf{U}^*. \tag{2.12}$$

Note that $\mathbf{V}$ and $\mathbf{U}$ are unitary, so that

$$\mathbf{A}\mathbf{A}^- = \mathbf{U} \begin{pmatrix} \mathbf{I} & 0 \\ 0 & 0 \end{pmatrix} \mathbf{U}^*.$$

Note that in general, $\mathbf{A}\mathbf{A}^- \neq \mathbf{I}$, unless $\mathbf{A}$ is square and nonsingular.

The definition above can be generalized to any singular square matrix.

Finally, we make a simple connection between eigenvalues, singular values and singularity of a square matrix. Using the SVD of a given matrix $\mathbf{A}$, we can say that $\mathbf{A}$

[2] Common notations for the pseudo-inverse also include $\mathbf{A}^\dagger$ and $\mathbf{A}^+$.

is nonsingular if and only if $\mathbf{\Sigma}$ has nonzero diagonal elements, indeed $\mathbf{A}^{-1}$ exists if and only if $\mathbf{\Sigma}^{-1}$ exists. A similar consideration holds with respect to the eigenvalue decomposition.

2.9 Functions of Matrices

It is possible to define functions of matrices in analogy to the familiar function on the real and complex numbers; see, e.g., Horn and Johnson (1991) for a more detailed treatment of this topic. For a given square matrix $\mathbf{A}$, a matrix polynomial of degree k is defined as

$$p(\mathbf{A}) = \alpha_0 \mathbf{I} + \alpha_1 \mathbf{A} + \alpha_2 \mathbf{A}^2 + \ldots + \alpha_k \mathbf{A}^k, \tag{2.13}$$

where the scalar coefficients $\alpha_0, \ldots, \alpha_k$ can be real or complex. The polynomial $p(\mathbf{A})$ is a matrix and there is no ambiguity in its construction, as long as matrix powers are carried out with the matrix product rule. If $\mathbf{A}$ is a diagonal matrix, that is $\mathbf{A} = \mathrm{diag}(a_{1,1}, \ldots, a_{n,n})$, then it can be easily verified that $p(\mathbf{A}) = \mathrm{diag}(p(a_{1,1}), \ldots, p(a_{n,n}))$, that is, the polynomial is applied to the single diagonal entries (cf. Exercise 6). We stress that this is only true for diagonal matrices, when their dimension is greater than one. If $\mathbf{A}$ is diagonalizable, that is $\mathbf{A} = \mathbf{X\Lambda X}^{-1}$, then it is possible to write

$$p(\mathbf{A}) = p(\mathbf{X\Lambda X}^{-1}) = \mathbf{X} \begin{bmatrix} p(\lambda_1) & \cdots & 0 \\ \cdots & \ddots & 0 \\ 0 & \cdots & p(\lambda_n) \end{bmatrix} \mathbf{X}^{-1} = \mathbf{X} p(\mathbf{\Lambda}) \mathbf{X}^{-1},$$

where we have used the property that $p(\mathbf{XAX}^{-1}) = \mathbf{X}p(\mathbf{A})\mathbf{X}^{-1}$ (this can be easily deduced first for $\mathbf{A}^k$, for any $k > 0$, and then for $p(\mathbf{A})$ using (2.13); see also Exercise 6). The calculation is rather interesting if we replace the polynomial p with a more general function f, such as $\exp(x)$, $\ln(x)$, $\sqrt{x}$, etc. Assume that f is a smooth function at the eigenvalues of $\mathbf{A}$. Then, as before, for diagonalizable $\mathbf{A}$ we can write

$$f(\mathbf{A}) = f(\mathbf{X\Lambda X}^{-1}) = \mathbf{X} \begin{bmatrix} f(\lambda_1) & \cdots & 0 \\ \cdots & \ddots & 0 \\ 0 & \cdots & f(\lambda_n) \end{bmatrix} \mathbf{X}^{-1} = \mathbf{X} f(\mathbf{\Lambda}) \mathbf{X}^{-1}.$$

In general the definition of a function of a matrix can be made rigorous without resorting to the diagonalization of $\mathbf{A}$, so that the matrix is not needed to be diagonalizable. We will assume that the function and the matrix we will use are all sufficiently well-behaved that the above definition can be used without special care.

Exercises and Problems

1. Given the matrix $\mathbf{A} = \begin{pmatrix} 2 & -4 \\ 1 & 0 \end{pmatrix}$, and the vector $\mathbf{b}^T = (-2, 1)$, verify that the vector $\mathbf{x}^T = [1, 1]$ is the (unique) solution to the system $\mathbf{Ax} = \mathbf{b}$.
We need to check that the definition is satisfied. Indeed, we have
$$\mathbf{Ax} = \begin{pmatrix} 2(1) - 4(1) \\ 1(1) + 0(1) \end{pmatrix} = \begin{pmatrix} -2 \\ 1 \end{pmatrix} = \mathbf{b}.$$
Note that $\mathbf{A}$ *is nonsingular, since the first row of the matrix is not a multiple of the second row (this is a sufficient consideration only in* $\mathbb{R}^2$*). Therefore the system solution is unique.*

2. Given the matrix $\mathbf{A} = \begin{pmatrix} 3 & -4 \\ -1 & 1 \end{pmatrix}$, verify that $\mathbf{x}^T = (-1 - \sqrt{5}, 1)$ and $\lambda = 2 + \sqrt{5}$ are respectively an eigenvector and the associated eigenvalue of $\mathbf{A}$.
We need to check that the definition is satisfied. Indeed, we have
$$\mathbf{Ax} = \begin{pmatrix} 3(-1 - \sqrt{5}) - 4(1) \\ -1(-1 - \sqrt{5}) + 1 \end{pmatrix} = \begin{pmatrix} -7 - 3\sqrt{5} \\ 2 + \sqrt{5} \end{pmatrix} \quad \textit{and} \quad \lambda\mathbf{x} = \begin{pmatrix} -7 - 3\sqrt{5} \\ 2 + \sqrt{5} \end{pmatrix}.$$

3. Show that the eigenvalues of an $n \times n$ real triangular matrix $\mathbf{A}$ coincide with its diagonal entries.
This can be checked by explicitly writing $det(\mathbf{A} - \lambda \boldsymbol{I}) = 0$. *Indeed, we have* $det(\mathbf{A} - \lambda \boldsymbol{I}) = (\lambda - a_{1,1})(\lambda - a_{2,2}) \cdot \ldots \cdot (\lambda - a_{n,n}) = 0$, *which is satisfied for* $\lambda = a_{i,i}$, $i = 1, \ldots, n$.

4. Show that the matrix $\mathbf{A} = \begin{pmatrix} 2 & 1 \\ 0 & 2 \end{pmatrix}$ only has one linearly independent eigenvector.
The matrix is triangular, therefore the eigenvalues are the diagonal elements (see exercise above). Hence, $\lambda_1 = \lambda_2 = 2$. *Using the definition* $\mathbf{Ax} = \lambda\mathbf{x}$, *eigenvectors of* $\mathbf{A}$ *are obtained by solving the singular system* $(\mathbf{A} - \lambda \boldsymbol{I})\mathbf{x} = \boldsymbol{0}$ *with* $\lambda = 2$. *We have*
$$(\mathbf{A} - \lambda \boldsymbol{I})\mathbf{x} = \begin{pmatrix} 0 & 1 \\ 0 & 0 \end{pmatrix}\begin{pmatrix} x_1 \\ x_2 \end{pmatrix} = \begin{pmatrix} 0 \\ 0 \end{pmatrix}$$
whose solution is $\mathbf{x} = (x_1, 0)^T$, $x_1 \in \mathbb{R}$. *No other linearly independent solutions exist.*

5. Show that the matrix $\mathbf{A} = \begin{pmatrix} 2 & 1 & 0 \\ 0 & 2 & 0 \\ 0 & 0 & 2 \end{pmatrix}$ has two linearly independent eigenvectors.
Proceeding as above, one finds that $\lambda_1 = \lambda_2 = \lambda_3 = 2$, *and there are two linearly independent eigenvectors,* $\mathbf{x} = (x_1, 0, 0)^T$ *and* $\mathbf{y} = (0, 0, y_3)^T$, $x_1, y_3 \in \mathbb{R}$.

6. Show that if $\mathbf{A}$ is diagonal, $\mathbf{A} = \text{diag}(a_{1,1}, \ldots, a_{n,n})$, then $p(\mathbf{A}) = \text{diag}(p(a_{1,1}), \ldots, p(a_{n,n}))$ for any polynomial p.
The result follows from observing that for any $k \geq 0$, $\mathbf{A}^k = diag(a_{1,1}^k, \ldots, a_{n,n}^k)$.
7. Given a square diagonalizable matrix $\mathbf{A} = \mathbf{X}\mathbf{\Lambda}\mathbf{X}^{-1}$, show that $p(\mathbf{A}) = \mathbf{X}p(\mathbf{\Lambda})\mathbf{X}^{-1}$.
We write $p(A) = \alpha_0 I + \alpha_1 A + \cdots \alpha_k A^k$. We have $A^2 = AA = X\Lambda X^{-1} X\Lambda X^{-1} = X\Lambda^2 X^{-1}$. This in fact holds for any j, that is $A^j = X\Lambda^j X^{-1}$. Therefore,

$$p(A) = \alpha_0 XX^{-1} + \alpha_1 X\Lambda X^{-1} + \cdots + \alpha_k X\Lambda^k X^{-1} = Xp(\Lambda)X^{-1},$$

where in the last equality the matrices X and X^{-1} have been collected on both sides.

Chapter 3
Basic Statistical Concepts

3.1 Introduction

A key scientific challenge is to better understand the functioning of the environment. Informed analysis of observations can make a strong contribution to this goal. The most insightful analysis requires knowledge of the relevant environmental processes and of statistical methodologies, that can lead the analyst towards a true understanding.

Compared to other aspects of the environment, climatology has a rich archive of direct observations. This has created an opportunity for the application of a wide range of statistical methods. This chapter reviews some of the basic statistical concepts that have been applied to better understand climate processes and to represent physically based predictability in the climate system. Like many environmental datasets, climate observations are sampling processes that evolve in space and time; the analysis of spatial patterns in time series of fields is the core of this book. Most reference is made to the application of special statistical techniques to study the fluctuation of climate from year to year. An additional special challenge is given by the size of the historical record. Typically, an analyst is faced with about 30–40 years of reliable data, which is sufficiently long to tease out some clues about the functioning of the climate system, but sufficiently short to lend itself to considerable imaginative interpretation. Thus, it becomes important to have a good appreciation for the effective sample size, so as to apportion the appropriate weight to the result in the overall investigation. When estimated properly, statistical significance allows us to have the correct degree of surprise at the statistical outcome, and therefore allows us to give the correct weight to this clue in our attempt to understand the big picture.

3.2 Climate Datasets

Climate observations were traditionally made at a known location. On land, this would be a climate station; over the ocean, this would normally be a ship, such that the exact location of the observations needed to be reported in addition to the climate state. The raw climate datasets from satellites can take a different form,

A. Navarra and V. Simoncini, *A Guide to Empirical Orthogonal Functions for Climate Data Analysis*, DOI 10.1007/978-90-481-3702-2_3,

being samples of space–time averages across the domain covered by the satellite. The blending of traditional and satellite datasets is therefore a critical step. In the context of climate analysis, data usually represent distinct observations at different times, possibly but not necessarily obtained at constant time intervals. In this case the term time series is usually employed for the data.

Since climate evolves in a continuous field, climate observations are often interpolated to a regular grid before analysis. These interpolation schemes may take in mind the processes and scales in the physical environment, and the ability of the data to resolve those scales. A good example are datasets of SST (monthly mean Sea Surface Temperatures). More recently, physically based interpolation schemes have been used to generate complete fields that are dynamically and physically consistent. These datasets have become known as the *reanalysis datasets*. They represent an ambitious advance in the creation of environmental datasets. In many ways, the user of such datasets needs more than ever to be aware of the types of data that were used in the study. Yet with careful analysis, they can provide an extremely powerful tool to deepen the understanding of the climate system.

The family of methods that are described in the following chapters are often applied to gridded datasets, such that the vectors derived from the analysis can be plotted as spatial patterns. However, there is no need to restrict the analysis to the gridded datasets. Analysis can equally be made of individual station time series. If the network of stations is sufficiently dense, contours of the weights can again be constructed to better communicate the meaning of the derived pattern. Alternatively, regional indices of climate, or regional indices of other environmental indicators can be used.

3.3 The Sample and the Population

An important concept in statistics is the relation between sample and population. Applying this concept to the analysis of short environmental series is not straightforward. It is assumed that the sample is taken from an infinite size population. The challenge is to infer characteristics of the population from the sample. The problem for climate science is that most properties of the system are not stationary. The problems of decadal climate variability have been mentioned above. In addition, the relationship between two variables need not be stationary. It can depend on the background climate state that prevailed over the analysis period. In fact, the degree of association between two variables may actually have varied during the 30 year period itself – though the sample size will likely be too small to deduce with any certainty that a real change took place. Let us pause to ask what we would mean by "a real change". Assume that we find a run of 10 years when the correlation is lower than during the whole historical record. What we want to know is the following: in case the interannual variability were repeatedly run with the prevailing background climate state of those 10 years, would that low correlation be maintained? or, would the 10 years of low correlation be merely due to the inevitable sampling fluctuations that occur even when the correlation between two variables is statistically stationary?

In addition, how does the situation change when we take a sample of a correlation coefficient over 30 years? The question we are trying to answer by taking that sample is: if the same background climate state were to continually operate and generate an infinite number of years of interannual variability, what would the correlation between the two variables be? In other words, the population is an imaginary infinite set of realizations generated from a given background climate state. For the purposes of making inferences (see statistical significance section below), we must assume that the correlation coefficient was stationary over the 30 year period itself.

3.4 Estimating the Mean State and Variance

A critical step in climate analysis is nearly always the estimation of the background mean state. Given the data $x_1, \ldots, x_n$, the mean, or average, is given by

$$\bar{x} = \frac{1}{n}\sum_{i=1}^{n} x_i.$$

The computation of the mean is crucial to allow estimation of climate anomalies, given by the deviation from the mean, that is,

$$x_i' = x_i - \bar{x}.$$

The climate anomaly represents the departure from the assumed population mean at a given time for a given time series. If there is a systematic bias in the estimation of the mean from one location to another, this can introduce bias in the covariance of anomalies between the two series; see later chapters for a more detailed discussion. Most widely available datasets have given careful consideration to the estimation of the mean from which anomalies are calculated.

For the background mean state for a dataset, the dataset creator will have considered such features as the period with best data coverage. If one is working with the subsequent anomaly dataset, one still has to make a choice over which years to run your analysis. This requires careful consideration and some experimentation, because of multi-year (decadal and beyond) variability in the climate system. Choice of period can greatly impact the amount of variance represented by a decadal mode of variation. For example, an analysis over West Africa for 1971–2000 contains little decadal variability, whereas 1950–1980 is dominated by a decadal fluctuation.

The sample variance of the observed data is defined as

$$s^2 = \frac{1}{n-1}\sum_{i=1}^{n}(x_i - \bar{x})^2.$$

In particular, we note the multiplicative factor $\frac{1}{n-1}$, as opposed to the more intuitive factor $\frac{1}{n}$. The new factor allows the variance defined above to represent an unbiased estimator of the population variance; see Clarke and Cooke (1998). To intuitively explain this fact, we note that in s^2 there are $n - 1$ degrees of freedom, rather than n, because $x_1, \ldots, x_n$ are related via the mean $\bar{x}$. Therefore, roughly speaking, the division by $n - 1$ takes into account the actual number of degrees of freedom in the data.

The variance provides a measure of dispersion of data around the mean. The larger the variance, the more spread the data. It is important to remark that the variance is expressed in the square of the data measure unit. For this reason, its square root s, called the standard deviation, is also referenced. Both statistics introduced above are dimensional quantities. To be able to perform a meaningful comparison among data expressed in different measure units, observational data are usually standardized to adimensional numbers. This is achieved by using the following transformation:

$$z_i = \frac{x_i - \bar{x}}{s}. \tag{3.1}$$

The standardized variable has mean zero and standard deviation equal to one.

Exercises and Problems

1. Given the data $\{1.2, -1.0, 1.1, 0.8, -0.4, 0.95, -0.2\}$, determine their mean, variance and standard deviation. Then, standardize the variables by means of (3.1).
   We have $n = 7$. Simple computation gives $\bar{x} = 0.35$, $s^2 = 0.75583$ and $s = 0.86939$. Standardization using (3.1) provides the following new data (final results rounded to the first five decimal digits),

   $$\{0.97770, -1.5528, 0.86268, 0.51761, -0.86268, 0.69014, -0.63263\},$$

   for which we obtain $\bar{z} = 0$ and $s(z) = 1$.
2. Given the data $\{1.2, -19, 2.68, 0.8 - 3.0, 20.0, -0.2\}$, compute mean, variance, standard deviation. Compare the results with those of the previous exercise.
   We have $n = 7$. Simple computation gives $\bar{x} = 0.3542$, $s^2 = 129.7$ and $s = 11.39$. Although the mean is basically the same as for the previous data, the variance and the standard deviation are much larger in this case. This shows that these data are more spread around the mean, as it can be clearly noticed by directly inspecting the data.

3.5 Associations Between Time Series

The basis for applying EOF methods derives from the realization that the evolution of climate processes in time leads to relationships between time series of different atmospheric variables at nearby and remote locations. There are many ways to measure the nature and extent of a relationship between two time series. One of the most common is the Pearson correlation coefficient. This is closely related to the concept of least squares linear regression. To illustrate this concept, we first do the simplest thing possible to explore the relationship between two time series – we make a scatter plot of the observation pairs (x_i, y_i) (see the symbols "$*$" in Fig. 3.1).

Making an assumption of a linear relationship, we try to draw a straight line through the data points. We can fit the line to minimize the sum of squared errors in the Y variable. This line captures some of the variance in the independent series. In mathematical terms, this line yields the "best approximation" straight line, in the least squares sense, and it is given by the equation $y = b_1 x + b_0$, with

$$b_1 := \frac{\sum\limits_i (x_i - \bar{x})(y_i - \bar{y})}{\sum\limits_i (x_i - \bar{x})^2}, \qquad b_0 = \bar{y} - b_1 \bar{x}.$$

The fraction of variance represented, corresponds to the degree of association (cf. Fig. 3.1). Analogously, the line $x = c_1 y + c_0$ can be drawn to minimize the sum of squared errors in the X variable. The fraction of variance explained is the same as for the Y variable. The combination of the two coefficients b_1 and c_1 yields the

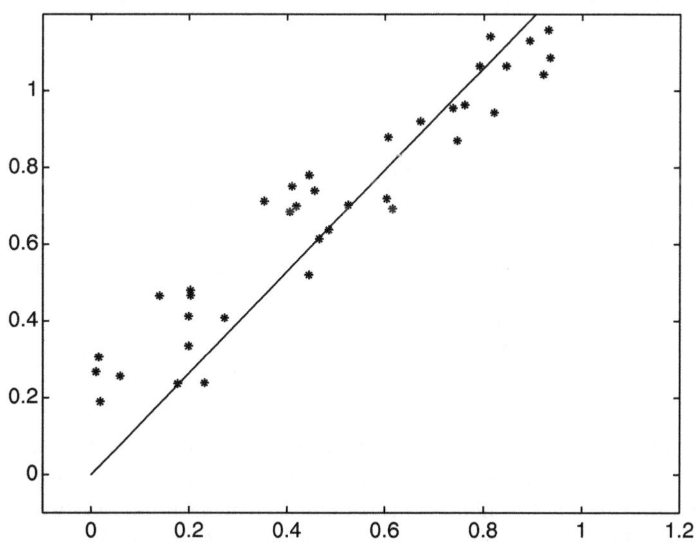

Fig. 3.1 Scatter plot of observations and fitting line

correlation coefficient r, which provides a measure of association among the two variables, and it is defined as

$$r^2 = b_1 \cdot c_1 = \frac{\left(\sum_i (x_i - \bar{x})(y_i - \bar{y})\right)^2}{\sum_i (y_i - \bar{y})^2 \sum_i (x_i - \bar{x})^2}.$$

To generalize this concept to multidimensional data, assume now that a set of $m \times n$ data $x_{1,1}, x_{1,2}, \ldots, x_{1,n}, \ldots, x_{m,n}$ is given. Here we are considering m variables and n observations (time series of length n for each of the m variables). Let $\bar{x}_j, \bar{x}_k$ be the means associated with the time series j and k. Analogously, we define the standard deviations s_j, s_k. For each pair of variables, the associated correlation coefficient is given by

$$r_{j,k} = \frac{1}{n-1} \sum_{i=1}^{n} \frac{(x_{j,i} - \bar{x}_j)(x_{k,i} - \bar{x}_k)}{s_j s_k}.$$

For the ith observation, $i = 1, \ldots, n$, the sum above multiplies the standardized jth and kth variables. The coefficient associated with these two variables is small (large) in absolute value, if both standardized variables are small (large), in all n observations. The normalization operates such that the correlation takes values between -1 (all points would lie on a backward sloping line) and 1 (all points would fall on a forward sloping line, cf. Fig. 3.1). Note that $r_{j,j} = 1$ for all j. In case standardization is not used, a related measure of association between deviations is the covariance coefficient, which can be viewed as a non-normalized correlation:

$$s_{j,k} = \frac{1}{n-1} \sum_{i=1}^{n} (x_{j,i} - \bar{x}_j)(x_{k,i} - \bar{x}_k). \tag{3.2}$$

Here $s_{j,j} = s_j^2$ is the variance of the jth variable. The matrix $\mathbf{S} = (s_{j,k})$ of all coefficients above is called the (cross-)covariance matrix and is symmetric, that is the covariance between the jth and kth variables is the same as the covariance between the kth and jth variables. The total variance of the field is then given by

$$T = \frac{1}{n-1} \sum_{i=1}^{m} \sum_{j=1}^{n} (x_{i,j} - \bar{x}_i)^2 = \sum_{i=1}^{m} s_{ii} = \text{trace}(\mathbf{S}), \tag{3.3}$$

showing that the total field variance is just the trace of the covariance matrix.

Both the above are related to the squared error departures from a linear relation. There are other ways to measure association. An example is the rank order Spearman correlation coefficient; see, e.g., Clarke and Cooke (1998).

Other measures could be determined based on absolute error. For instance, the Linear Error in Probability Space (LEPS) works on the mean absolute difference in the ranking, or cumulative probability. Associations could be measured in terms of the extent to which variance is explained by some specified non-linear relationship, such as quadratic or log linear. The correlation coefficients introduced above can be collected in one matrix, that more clearly visualizes the association of each time series with all others.

For instance, the correlation matrix is given by

$$\mathbf{R} = \begin{pmatrix} r_{1,1} & r_{1,2} & \cdots & r_{1,m} \\ r_{2,1} & r_{2,2} & \cdots & r_{2,m} \\ \vdots & \ddots & \ddots & \vdots \\ r_{m,1} & r_{m,2} & \cdots & r_{m,m} \end{pmatrix}.$$

As will be shown, there are special properties of correlation and covariance matrices that can be uncovered by a Principal Component Analysis. Matrix properties of other measures of association have not been as much investigated; they will be alluded to in the following chapters.

Exercises and Problems

1. Given the data $x = \{-1.1, 0.8, 1.2\}$ and $y = \{0.6, 0.9, 2.4\}$, determine the covariance and correlation matrices.
   We have $m = 2$ variables, and $n = 3$ observations. Therefore, $\bar{x} = 0.3$, $\bar{y} = 1.3$, $s(x) = 1.2288$ and $s(y) = 0.9643$, so that the standardized variables are $z(x) = \{-1.1393, 0.40689, 0.73241\}$ and $z(y) = \{-0.72587, -0.41478, 1.1406\}$. The correlation coefficient is given by $r_{1,2} = 1/2(z(x)_1 z(y)_1 + z(x)_2 z(y)_2 + z(x)_3 z(y)_3) = 0.74$ (note that the computation of $r_{1,2}$ is actually done with full accuracy and only the first 2 decimals are reported). Hence, the corresponding matrix is

   $$\boldsymbol{R} = \begin{pmatrix} 1 & 0.74 \\ 0.74 & 1 \end{pmatrix}.$$

   The value of $r_{1,2}$ shows a significant positive correlation between the two variables. Analogously, the covariance is given by $s_{1,2} = 0.885$.
2. Given the data $x = \{-1.1, 0.8, 1.2\}$, $y = \{0.6, 0.9, 2.4\}$ and $z = \{4.2, -1.1, 6.8\}$, determine the covariance and correlation matrices.
   We have $m = 3$ variables, and $n = 3$ observations. The first two sets are as in the previous example. We have, $z = 3.3$, $s(z) = 4.0262$, so that the new standardized variable is $z(x) = \{0.22354, -1.0929, 0.86931\}$. We obtain $r_{1,3} = -0.00313$ and $r_{2,3} = -0.64$. The correlation between the y and z variables is significant, whereas that between x and z is negligible. Analogously, we obtain $s_{3,3} = 16.21$, $s_{1,3} = -0.155$ and $s_{2,3} = 2.49$.

3. Given the data $x = \{-1.1, 0.8, 1.2\}$, $y = \{0.6, 0.9, 2.4\}$ and $z = \{104.2, -100.1, 126.8\}$, determine the covariance coefficients. Comment on the role of dimensionality.
Only the third variable has changed. We have $s_{3,3} = 1562.2$, $s_{1,3} = -40.90$ *and* $s_{2,3} = 53.29$. *Note that the larger variability is due to the significantly different unit of z, which is also reflected in the covariance coefficients.*

3.6 Hypothesis Testing

In an attempt to infer conclusions on an unobservable population, we can set about estimating the extent to which our available sample can tell us something about that population. Let us consider the simple example of testing whether the population mean is zero. Statistical significance is estimated by formally expressing two possibilities that we need to choose between. Here, the first one is that the population mean is zero. As an alternative possibility, we can say that the population mean is not zero (other options may be more significant in some cases, such as "mean greater than zero"). Formally, the original hypothesis is termed the null hypothesis (H_0), whereas the second one is called the alternative hypothesis (H_1), and these are written as

$$H_0 : \quad \mu = 0, \qquad H_1 : \quad \mu \neq 0.$$

We want to distinguish between these two possibilities in a way that allows us to know the likelihood that our choice is in fact wrong (i.e. how surprised we should be if our decision turns out to be the wrong one). We start out by assuming that H_0 is true. If H_0 is true, then the sample should obey certain statistical properties. If the sample does not reflect these properties, then we start to doubt H_0. For example, we can define a test statistic whose distribution we know under the assumption that H_0 is true and we explore to what extent our sample obeys this distribution.

A particularly popular distribution is the *normal* distribution, as it represents an effective model for data stemming from a variety of applications. Data following a normal distribution distribute around their mean with a probability that decreases significantly as data move away from the mean. The set of normally distributed variables with mean μ and variance σ^2 is usually denoted by $N(\mu, \sigma^2)$. The probability of normal data distributes along a bell-shaped curve, as described in the plots of Fig. 3.2 for various values of μ and σ. In other words, the probability that a sample taken from an $N(\mu, \sigma^2)$ normal population has mean in the interval $[\mu - d, \mu + d]$ equals the area of the region below the bell-shaped curve, with extremes on the ascissa at $\mu - d$ and $\mu + d$. A normally distributed variable x with mean μ and variance σ^2 can be transformed into a standardized normally distributed variable in $N(0, 1)$ by means of the change of variable $z = (x - \mu)/\sigma$. Reference values for a variable z in $N(0, 1)$ are tabulated and can be used for hypothesis tests. Most

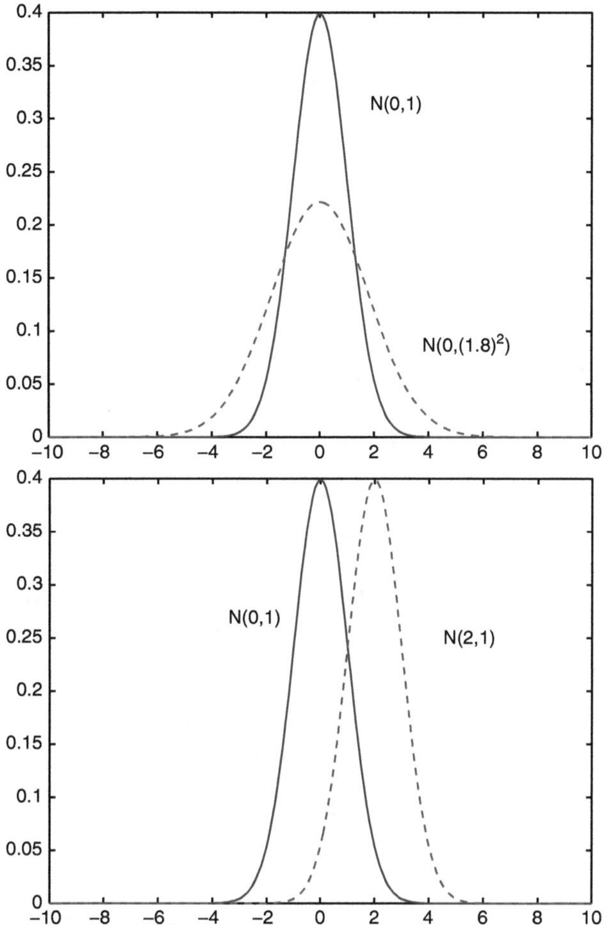

Fig. 3.2 Normal distributions for various values of μ and σ

statistical computer software provides a pretty accurate evaluation of the probability and other quantities associated with the normal distribution.

The trick in hypothesis testing is to define powerful test statistics, such as the standardized statistic

$$z = \frac{x - \mu}{se},$$

where se is the standard error of x, given by

$$se = \frac{\sigma}{\sqrt{n}},$$

σ is the population standard deviation and n is the sample size. The standard error represents the standard deviation of the sample mean distribution. In other words, imagine estimating the mean of the population ten times, each time taking a sample of n individuals from the population. The ten resulting sample means will all be slightly different. The expected standard deviation of the sampled means is what we refer to as the standard error of the estimated mean.

Assume that H_0 is true. If the sample is sufficiently large, namely if n is sufficiently large, then z approximately behaves as if it were normally distributed with mean 0 and standard deviation 1.

Now, if the distribution of z departs substantially from 0, then we may start to doubt H_0. If the variable z were exactly normally distributed with zero mean and unit variance, then only on 2.5% of occasions z would take a value of $+1.96$ or higher. Likewise, on a further 2.5% of occasions, z would take a value not greater than -1.96 (cf. Fig. 3.3). That is, there is a 5% chance of the absolute value of z being greater than 1.96. So, if z takes an absolute value greater than 1.96, such a result is certainly quite surprising if in fact, the true mean is zero (surprising because we only expect it to happen on 5% of occasions when we sample a population mean with mean 0). Note that we can never be certain that H_0 is wrong. For statistical significance, we may decide that something that would only happen by chance on 5% of occasions is just too surprising, and that the wisest choice to make in this situation is to conclude that the available evidence does not support H_0. That is, at the 5% level of significance, we reject H_0 and accept the alternative hypothesis H_1, that the mean is not equal to zero. However, in terms of acquiring clues about the overall functioning of the environment, we may prefer not to work in the discrete terms of rejection or acceptance of H_0. Rather, acknowledging that using statistics alone, we can never distinguish between the two hypotheses with certainty, we may prefer to note the likelihood that H_0 can be rejected based on statistics alone, and absorb this information into broader evidence based on physical theories and physically based models.

The above approach estimates the probability of rejecting H_0 by starting with the assumption that H_0 is actually true. This is the usual way to frame a statistical

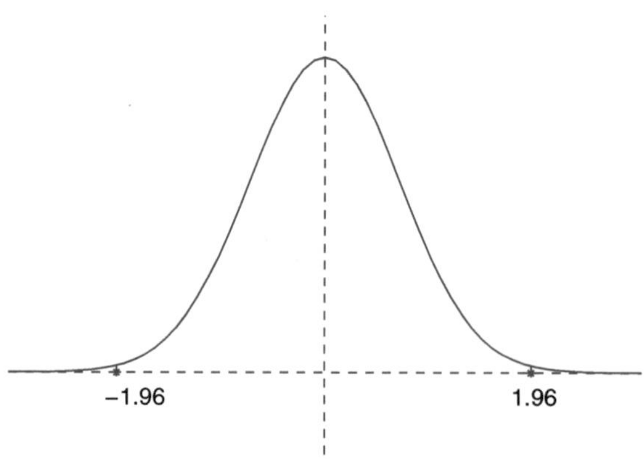

Fig. 3.3 Normal distribution. The area of the region below the curve and absissas in $[-1.96, 1.96]$ is equal to 0.95

significance test, on the premise that the analyst is usually interested in whether H_0 can be rejected, such as with whether a correlation coefficient is non-zero. Here, if we choose to reject the idea that the true correlation is zero, what is the probability that we are wrong (and in fact, there is a linear association between the two variables). This probability of wrongly rejecting H_0 is often termed the probability of making a Type I error, and is the statistical significance probability, alpha. However, there is another error type that can be made, usually referred to as Type II error, that of accepting H_0 when in fact H_0 should be rejected. This probability can also be estimated assuming the distribution of test statistics. However, it is generally not considered as useful as the Type I error probability, that focuses on whether we can reject H_0.

The distribution of the reference statistic z is easy to derive and work with. In many instances the test statistic is more complex. A typical complication appears when the standard deviation of the population is not known (of course, this is usually the situation we find ourselves in). In this situation, we can use the Student statistics, or t-statistic, in which the population standard deviation is replaced by the sample standard deviation, that is

$$t_0 = \frac{x - \mu}{\bar{s}} \sqrt{n} \tag{3.4}$$

The new variable t_0 depends on n, more precisely on $n - 1$, and for each value of n, t_0 follows a specific distribution. Is it important to stress that to be able to employ the Student distribution as test statistic, we need to assume that the given sample comes from a normal distribution.

As n grows, the Student distribution increasingly resembles the normal distribution. The likelihood of t_0 exceeding a reference value is tabulated, for different values of $n - 1$, called the degrees of freedom, Df, which is related to the size of the available sample. The degrees of freedom is a complicated issue for many climate analyses. The above holds if each term in the sample is independent. However, in many climate time series, adjacent observations are correlated in time, and this reduces the effective degrees of freedom (and can complicate the distribution of the test statistic). This is particularly a challenge for estimating the significance of the relationship between two variables. The correlation coefficient significance is very difficult to estimate because of this effect; see von Storch and Zwiers (1999). This problem transfers into the estimation of significance for EOFs, since they themselves are summaries of the cross-correlations/covariances in datasets.

Exercises and Problems

1. Assume that a sample of 100 units is taken from a population which was in the past known to have mean $\mu = 12.3$ and standard deviation $\sigma = 15$. The computed sample mean is $x = 14.2$. Carry out a hypothesis test with 5% level of significance, to analyze whether the population mean has changed.

We set H_0*:* $\mu = 12.3$ *and* $H_1 : \mu \neq 12.3$. *We have* $z = (x - \mu)/\sigma = 0.12$. *The critical region for 5% level of significance would be* $|z| > 1.96$, *therefore the new variable z is well away from the critical region. We do not reject the null hypothesis.*

2. What would happen in the example above if the standard deviation were $\sigma = 0.9$? What if the significance level were 1%?
   With the same framework as before, we have $z = (x - \mu)/\sigma = 2.11$, *hence this variable falls within the critical region* $|z| > 1.96$. *We have to reject the null hypothesis in favor of the alternative hypothesis* H_1 *for a 5% level of significance. For a significance level equal to 1%, the corresponding critical region is* $|z| > 2.57$, *so that the null hypothesis would not be rejected.*

The inherent difficulty associated with the effective number of degrees of freedom in the Student statistics is one of the reasons why alternatives such as Monte Carlo estimates of significance are attractive. To illustrate the concept, consider that we have two time series of length 30 years. Each time series has serial correlation and can be represented by an autoregressive process:

$$x_t = a x_{t-1} + z_t . \tag{3.5}$$

We can use random number generators in combination with the above model to simulate 500 pairs of time series with the same serial correlation properties as the original two series. The distribution of the 500 correlations between each randomly generated pair of series is now constructed empirically. We expect the mean of the correlations calculated to be zero, but the spread will depend on the degree of autocorrelation in the two series. If the pair of series are highly auto-correlated, the location of the correlation magnitude that occurs on 5% of occasions will be much higher than if the pair of series were uncorrelated. Now we are using the correlation itself as the test statistic, knowing the distribution of the sample correlations under the assumption that the true population correlation is 0. The correlation magnitude corresponding to the 5% significance level can be found by identifying the threshold above which were found only 5% of the sample correlations. The temporal d.f. problem is also present for methods devised to estimate the statistical significance of EOFs. Higher percentage of variance explained are expected by chance, when time series used in the EOF analysis contain serial correlation. Thus caution is needed not to place excessive weight on significance estimates of EOFs when series have serial correlation.

3.7 Missing Data

Dealing with missing data is an important aspect for application of EOF methods. In some datasets, the fields will have been made complete for the analyst, in which case the analyst should investigate carefully the way the data were interpolated and possible consequences for EOF analysis as discussed below. The EOF methods are applied to correlation and covariance matrices. It is tempting to calculate correlations using the available data for each pair of series, assuming that this gives the best estimate of the correlation between each series, even if some correlations are based on a smaller sample than others. However, this approach can lead to problems with the inversion of the correlation/covariance matrices to derive the EOF solutions. It is usually best to make all series complete in some way over the analysis period.

Usually, the analyst decides on a fixed analysis period (say, 1961–1990) and decides on the maximum number of missing values that is acceptable for a series to be included (say, at least 25 out of 30 values must have data). A simple and quite robust solution to missing data is to set all missing values in a series equal to the mean of the available data for that series. This will ensure the missing values are all zero anomalies when the correlation/covariance matrices are calculated. Zero anomalies have least impact on the correlation/covariances. While it can reduce some genuine cross correlations between time series and this can distort the EOF solutions, it is nonetheless a cautious conservative approach and as such, is an attractive solution. Application of more sophisticated interpolation methods requires care for any increase in correlations/covariances that it may introduce into the datasets.

Chapter 4
Empirical Orthogonal Functions

4.1 Introduction

The atmospheric fields are three-dimensional fields by nature, the variation in longitude, latitude and altitude of winds, temperatures and the other quantities are normal. A major jump forward in the development of climate science was reached when it was realized that the analysis of simultaneous values of the variables contained significant information. Indeed, to advance scientific understanding it is essential to have a view of the relation that links together various climate variables, for instance the temperature and the pressure in various places.

The covariance matrix is the statistical object that rigorously describes such relations. Let us consider the case of monthly time series data, for instance temperatures, for stations in m locations and for a period of n months. Each station can be represented by a vector, $\mathbf{x}_i = (x_i(1), x_i(2), \ldots, x_i(n))$, and the stations can be arranged in an array;

$$\begin{array}{lclllll} \mathbf{x}_1 & = & [x_1(1) & x_1(2) & \ldots & x_1(n)] \\ \mathbf{x}_2 & = & [x_2(1) & x_2(2) & \ldots & x_2(n)] \\ \vdots & & \vdots & \vdots & \vdots & \vdots \\ \mathbf{x}_m & = & [x_m(1) & x_m(2) & \ldots & x_m(n)]. \end{array}$$

Organizing the data in this way suggests an alternative description for the global data set of station time series. Instead of ordering them as time series, we can order along the vertical columns. The array we obtain is equivalent to considering vectors of values at the same time, the so called synoptic view. In mathematical terms we can treat the array as an $m \times n$ matrix as follows

$$\mathbf{X} = \begin{bmatrix} x_1(1) & x_1(2) & \ldots & x_1(n) \\ x_2(1) & x_2(2) & \ldots & x_2(n) \\ \vdots & \vdots & \vdots & \vdots \\ x_m(1) & x_m(2) & \ldots & x_m(n) \end{bmatrix}, \qquad (4.1)$$

A. Navarra and V. Simoncini, *A Guide to Empirical Orthogonal Functions for Climate Data Analysis*, DOI 10.1007/978-90-481-3702-2_4,

or introducing the synoptic vectors $\mathbf{x}_j^* = [x_1(j), x_2(j), \ldots, x_m(j)]$ for $j = 1, 2, \ldots, n$ we have the matrix $\mathbf{X}$ defined in terms of column vectors,

$$\mathbf{X} = [\mathbf{x}_1, \mathbf{x}_2, ..., \mathbf{x}_n],$$

where the rows of the matrix now describe the values at the same spatial location.[1]

In the analysis of meteorological or climatological data it is very common that time series come from observations or from numerical simulations taken at regular intervals. A typical example are for instance temperatures taken at several stations around the world, grouped in monthly means, so that only one value per month is available. In this case the columns of the data matrix $\mathbf{X}$ indicate the time series at each station, whereas the synoptic vectors, i.e. the rows of the data matrix, describe the geographic distribution of the temperature at a particular time. The time evolution of the temperature can then be followed by looking at the sequence of geographical maps depicting the temperature distribution every month. Usually, for interpretation purposes it is useful to use contouring algorithms to represent the field as a smooth two-dimensional function. Contouring is not a trivial operation and especially for observations that are distributed in space with large gaps and far from being a regular covering of the surface of the Earth, must be done with care. Sophisticated techniques, known as data assimilation, are employed to make sure the data sets are put on regular grids in a physically consistent manner. In any case, either that we are looking at data from modelling, or that we are working with observations coming from data assimilation systems, we end up with data on regular grids, covering the Earth with a regular pattern.

As already mentioned, we use two data sets to illustrate our discussion. The first is a time series of monthly mean geopotential data at 500 mb (Z500), obtained from a simulation with a general circulation model forced by observed values of monthly mean Sea Surface Temperatures (SST). The data sets cover 34 years, corresponding to the calendar years 1961–1994. The Z500 set is a very good indicator of upper air flow, since the horizontal wind is predominantly aligned along the geopotential isolines. Figure 4.1 shows a few examples taken from the data set. It is possible to note the large variability from one month to the other (top panels), but also the large variability at the same point, as the time series for the entire series (lower panels) show. It is clear that the geopotential at 500 mb is characterized by intense variability in space and time and a typical month may be as different from the next month as another one chosen at random.

We can consider the maps at each month as a vector in a special vector space, the data space. Each vector in this data space represents a map, a possible case of a Z500 monthly mean. The space covers all possible shapes of the Z500, the vast majority of which will never be realized, like the one in which the heights are constant everywhere, or some other similar strange construction. The mathematical dimension of

[1] This matrix notation is very common in meteorological data, whereas in many other fields, data are stored as an $n \times m$ matrix, namely as $\mathbf{X}^*$. This difference affects the whole notation in later chapters, when defining the covariance matrix and other statistical quantities.

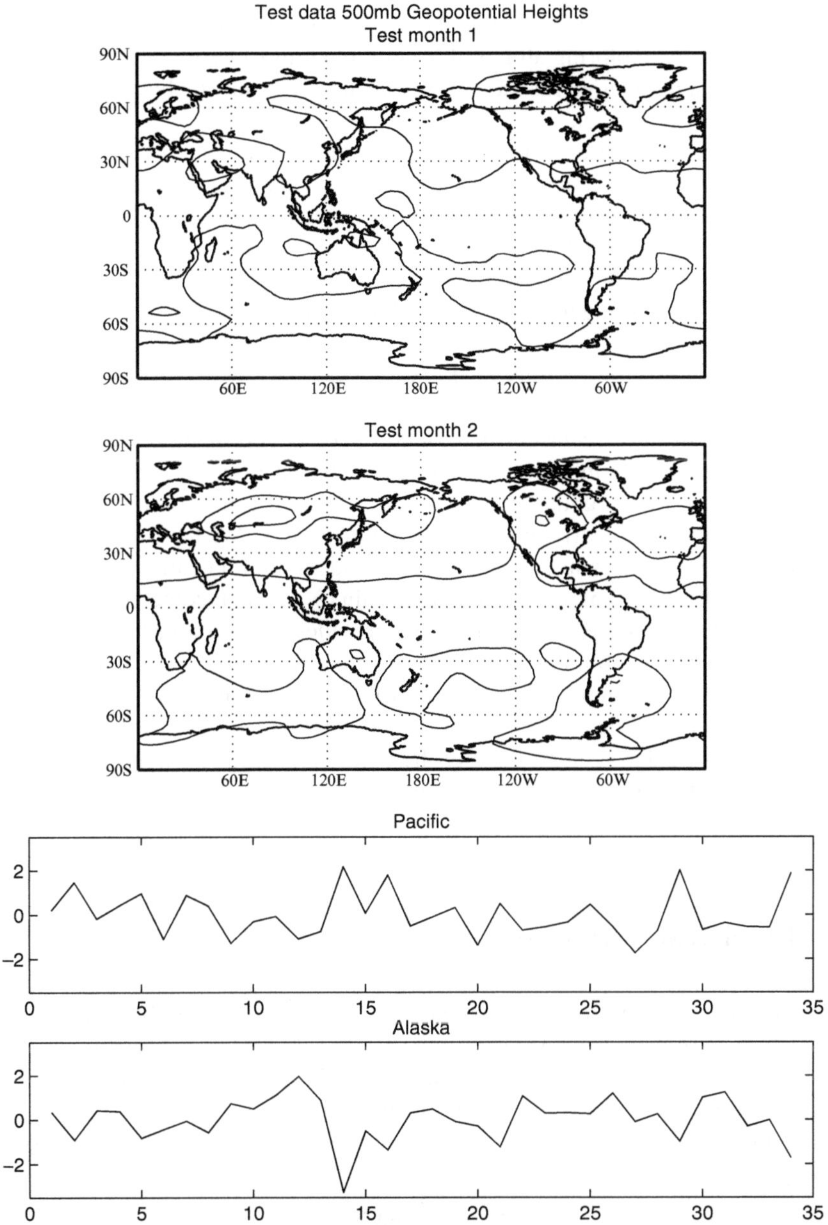

Fig. 4.1 Examples of Z500 data set. The maps are seasonal means for winter taken from a model simulation with prescribed Sea Surface Temperatures for a period of 34 years from 1961 to 1994. Here seasonal anomalies are shown, after removal of the climatological winter mean

the vector space is very high, it is equivalent to the number of observations points, n. In the case of the test data sets used in this book, which have 96 points along the longitude for each latitude line and 48 latitude lines, it is $n = 4608$. In fact, this value is so large that the question arises whether the geopotential in the natural variability of climate is really exploring all possible combinations of numbers in this 4608 observation points, in other words the data set we are using may be lying on a subspace of much smaller dimension.

One question that arises when one is confronted with such an intricate pattern of behaviors is whether there are special ways in which the Z500 fields can express themselves, in other words whether typical recurring patterns exist. In the following we introduce a possible technique to identify patterns of this kind.

We can gain some insight into the determination of possible subspaces that are frequently visited if we again consider the data matrix

$$\mathbf{X} = [\mathbf{x}_1, \mathbf{x}_2, ..., \mathbf{x}_n].$$

The simplest case of a recurrent pattern subspace occurs if the n vectors are not linearly independent. In this case, the successive realizations *1,...,* n are linear combinations of just a few fundamental patterns. The first thing to attempt is then to assess the number of linearly independent columns of $\mathbf{X}$. We have seen in Sect. 2.9 that the rank of $\mathbf{X}$, that is its number of independent columns, may be obtained by using the SVD. If the rank of $\mathbf{X}$ is less than min $\{m,n\}$, then some columns will be just a mixture of the others. The SVD also provides us with a basis for the vector space spanned by the columns of $\mathbf{X}$ (and also for the vector space spanned by the rows, but we are not interested in that part now) according to $\mathbf{X} = \mathbf{U\Sigma V}^T$. The decomposition of the data space gives us a mathematical basis for the maps, the right and left singular vectors, namely the columns of the two matrices $\mathbf{V}$ and $\mathbf{U}$. However, they do not seem to have any special meaning. This is what we will try to describe in the rest of this chapter.

4.2 Empirical Orthogonal Functions

In the discussion that follows we assume that the data matrix has been transformed so as to have zero mean vector. This can be easily achieved by subtracting the mean value to each corresponding row of $\mathbf{X}$, that is $\mathbf{X} = \mathbf{X}_{orig} - \bar{\mathbf{x}}\mathbf{1}^*$, where $\mathbf{1}$ is the column vector of all ones, while $\bar{\mathbf{x}}$ is the vector of sample means.

With these scaled data, the covariance matrix $\mathbf{S}$ can also be written in the following way

$$\mathbf{S} = \frac{1}{n-1}\mathbf{X}\mathbf{X}^T. \tag{4.2}$$

It is easy to check that the combination in (4.2) indeed satisfies the definition of covariance matrix in (3.2). We can then use the decomposition in singular values of $\mathbf{X}$ that we introduced in the previous section so that

$$\mathbf{S} = \mathbf{U\Sigma V}^T\mathbf{V\Sigma U}^T = \mathbf{U\Sigma}^2\mathbf{U}^T. \tag{4.3}$$

This expression reveals that the left singular vectors of the data matrix $\mathbf{X}$ are also the eigenvectors of the (symmetric) covariance matrix $\mathbf{S}$. We can now understand a little better the role of the vectors in $\mathbf{U}$. The number of independent maps in the data sets, the columns of $\mathbf{U}$, are the same as the eigenvectors of the covariance matrix. Since one is mostly concerned with these modes, which are invariant under scaling of $\mathbf{S}$, the factor $1/(n-1)$ is often omitted when using the covariance matrix $\mathbf{S}$. On the other hand, care should be taken in working with the eigenvalue diagonal matrix $\mathbf{\Sigma}^2$, since scaling $\mathbf{S}$ correspondingly scales the eigenvalues.

According to (4.3), the vectors in the unitary matrix $\mathbf{U}$ are such that the covariance matrix in that basis is diagonal, that is each vector $\mathbf{u}$ is uncorrelated with the others and contributes to the total variance an amount given by the diagonal element of $\mathbf{\Sigma}^2$; indeed, because of the invariance property of the trace, the total variance is also given by the sum of the squared singular values. Dividing them by the trace we can get the percentage contribution μ_i of each mode σ_i^2,

$$\mu_i = \frac{\sigma_i^2}{\sum_{i=1}^{n} \sigma_i^2}. \tag{4.4}$$

If $\mathbf{X}$ is not full rank, that is if some of its columns/rows are linearly dependent, then we will get fewer nonzero singular values, say q, and associated left singular vectors. Equivalently, the covariance matrix has $q < \min\{m, n\}$ nonzero eigenvalues with corresponding q orthonormal eigenvectors. The independent modes of variations in $\mathbf{U}$, associated with nonzero singular values, via the SVD of the data matrix, or via the eigenanalysis of the covariance matrix, are called Empirical Orthogonal Functions, or in short EOF. Note that in the wide literature on Principal Component Analysis, these modes are called *Principal Components*, or PC (see, e.g., Jolliffe 2002).

4.3 Computing the EOFs

The calculation of the eigenpairs of the covariance matrix can be a difficult numerical problem because the dimension of the matrix tends to grow with the number of observation points. A much faster way to obtain the EOF is to use (4.3) and perform an SVD on the data matrix. The difference can be of several orders of magnitude in terms of computational cost and it should be considered the right way to get the EOFs. The SVD decomposition is also more stable and accurate. EOFs can then be readily computed using MATLAB that has primitive functions for both the SVD and eigenvectors calculation, the former being preferred for computational efficiency (cf. Sect. 4.3.1). As an example, below is a Matlab code that generates *all* EOF and projected selected components.

```
function [u,lam,proj]=eoffast(z,indf,nmode,nproj)
%
% Compute all EOF of matrix z and expand it for nmode modes
% Also returns nproj projection coefficients
%
resol = [96 48];

[uu,ss,vv]=svd(z,0);            %Memory saving decomposition

lam = diag(ss.^2)/sum(diag(ss.^2)); % Explained variances
%
u=zeros([resol(1)*resol(2) nmode]);
u(indf,1:nmode)=uu(:,1:nmode);

proj=vv*ss(1:nproj,:)';                 % Compute projections

return
```

For a large matrix, the call to svd in the algorithm above may be replaced by a call to svds, which computes only a subset of all singular triplets of the given matrix, as requested by the user. Also in this case, this procedure should be preferred to the nowadays obsolete strategy of using the power method to compute a few of the largest eigenvalues of the correlation matrix.

4.3.1 EOF and Variance Explained

The plots in Fig. 4.2 show the results of performing EOF on a test data set constructed from the Z500 test set, but in which we have artificially restricted the variations to only three independent vectors. An entire data set has then been created by random combinations of the three. It is possible to see how the EOF has correctly identified that there are only three independent vectors.

Following the ratio in (4.4), the three basic modes are contributing to the variance a fraction that is given in the bottom of Fig. 4.2. The first mode is contributing more than 60% of the variance, whereas the remaining two are more or less equally dividing the rest. This is somewhat strange since the data set was constructed by combining the basic vectors with random coefficient uniformly distributed, we would expect each vector to contribute equally to the variance of the field. One hypothesis is that the sample size is too small, but we can see in the subsequent picture (Fig. 4.3) that increasing the sample size (top row) does not modify the distribution, and the contribution to the variance remains non-uniform.

The solution to the puzzle must be found in the fact that we have arbitrarily selected the basis vectors. Investigating the vectors it can be found that they are not mutually orthogonal since their scalar product is not zero. Removing the dependency with an orthogonalization procedure and repeating the analysis we obtain the bottom row of the figure. We can see that now each vector contributes uniformly to

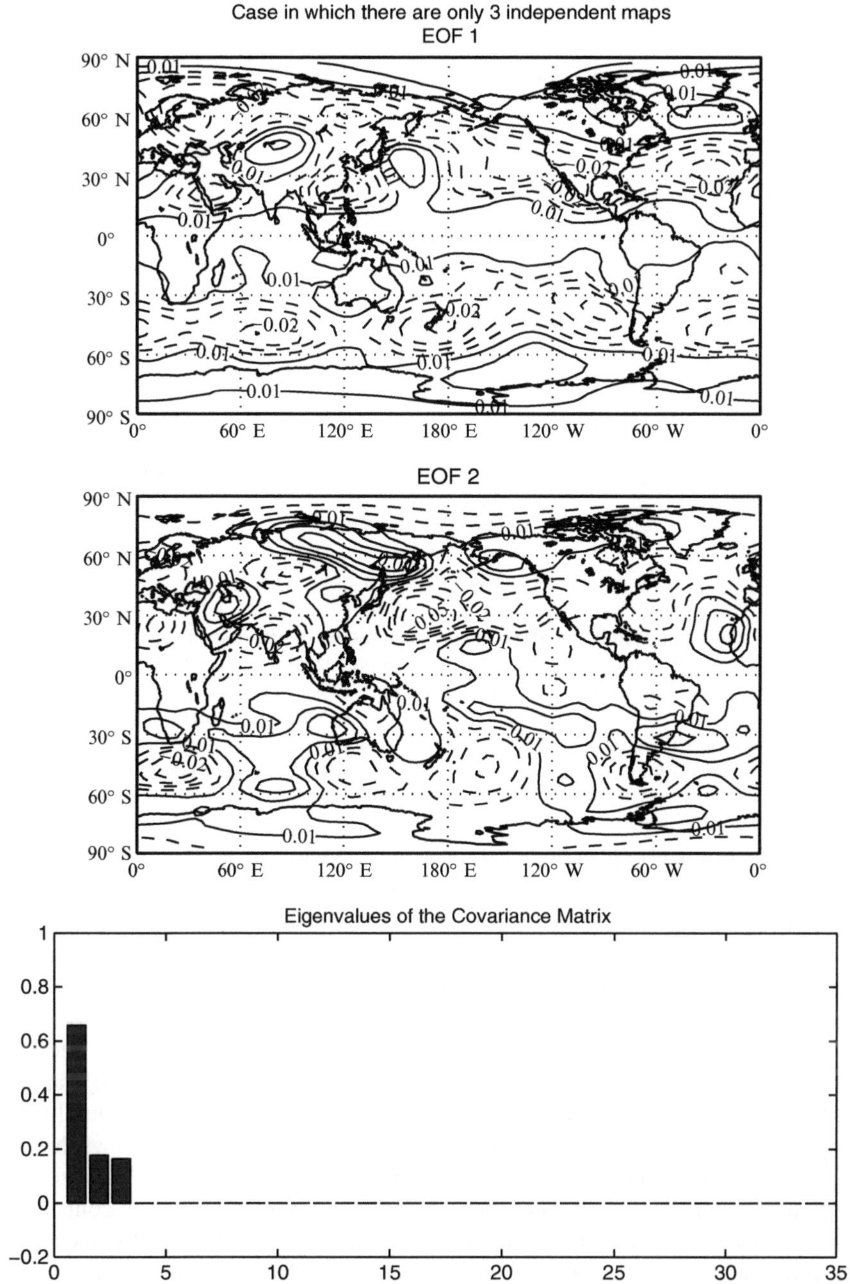

Fig. 4.2 Empirical Orthogonal Functions (EOF) for the test case in which only three independent maps have been chosen in the Z500 data set and then a complete 34 years set has been reconstructed with random combination of the three. Here are shown the first and the second mode (*upper panels*) and the spectra of the singular values (*lower panels*) that reveals that three modes have been correctly identified

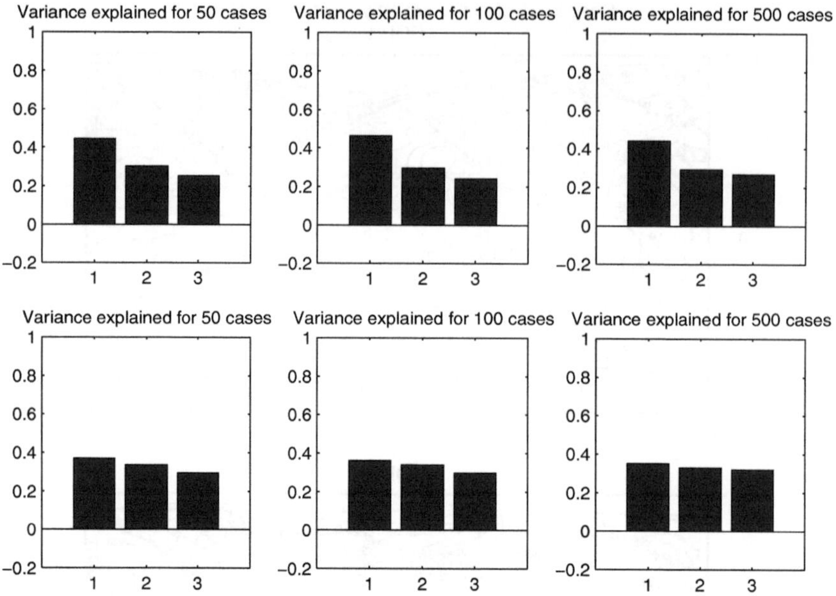

Fig. 4.3 Variance explained by the three independent modes, in the case of the preceding picture. The top row shows the variance explained as the sample size increases, passing from 50 to 100 and 500 global maps. The distribution is stable, in fact it is well captured even with 50 cases, but it is not uniform. In the bottom row we see the variance explained as a function of the sample size for a test data set in which the basic vector are orthonormalized prior to the generation of the data. The EOF correctly estimate an equal contribution to the variance, and a modest improvement of the accuracy can be seen with the increase of the sample size.

the variance as it was expected. In this case the estimation improves with the sample size as we pass from 50 to 500 cases. This simple experiment leads us to conclude that the EOF are abstract patterns, that can be used to better (more cheaply) represent the system total variance. Finally, it is worth noticing that had we used standardized data, or equivalently the correlation matrix, we would not have observed this intriguing phenomenon.

Figure 4.4 shows the EOF for the Z500 data set without any treatment. The mode patterns are qualitatively similar, but the spectrum is quite different, and it appears that there are no zero singular values, corresponding to all 34 months being linearly independent. The interpretation in terms of covariance eigenvectors shows that they are not all equally important, since we can now rank them according to the size of the contribution to the total variance. Some vectors give a relatively large contribution to the variance, whereas others are basically contributing nothing. This means that this field has a preference to vary according to the first modes of variations and consequently the corresponding patterns are most typical.

In general we will obtain as many significant EOF as the smallest number between the length of the time series n and the number of observation points m. With our test data sets and almost always when treating with climate or weather data, the number of observation points, either as grid points from simulations or station

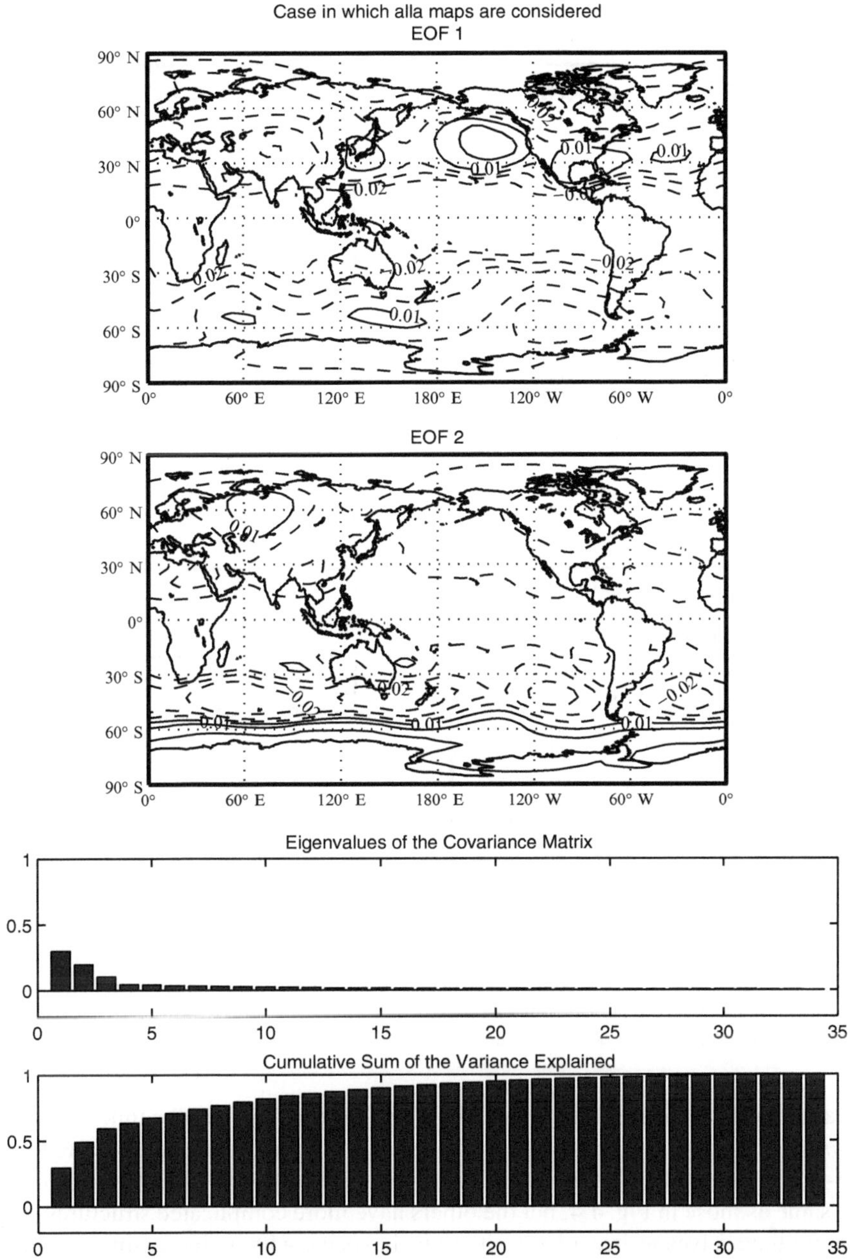

Fig. 4.4 Empirical Orthogonal Functions (EOF) for the Z500 data set. Here are shown the first and the second mode (*upper panels*) and the singular values (*middle panel*) that reveals that several independent modes have been correctly identified. For instance, the *bottom panel* illustrates the cumulative variance explained by the first k modes: the first five modes describe together about 60%. The last ten modes have zero contribution

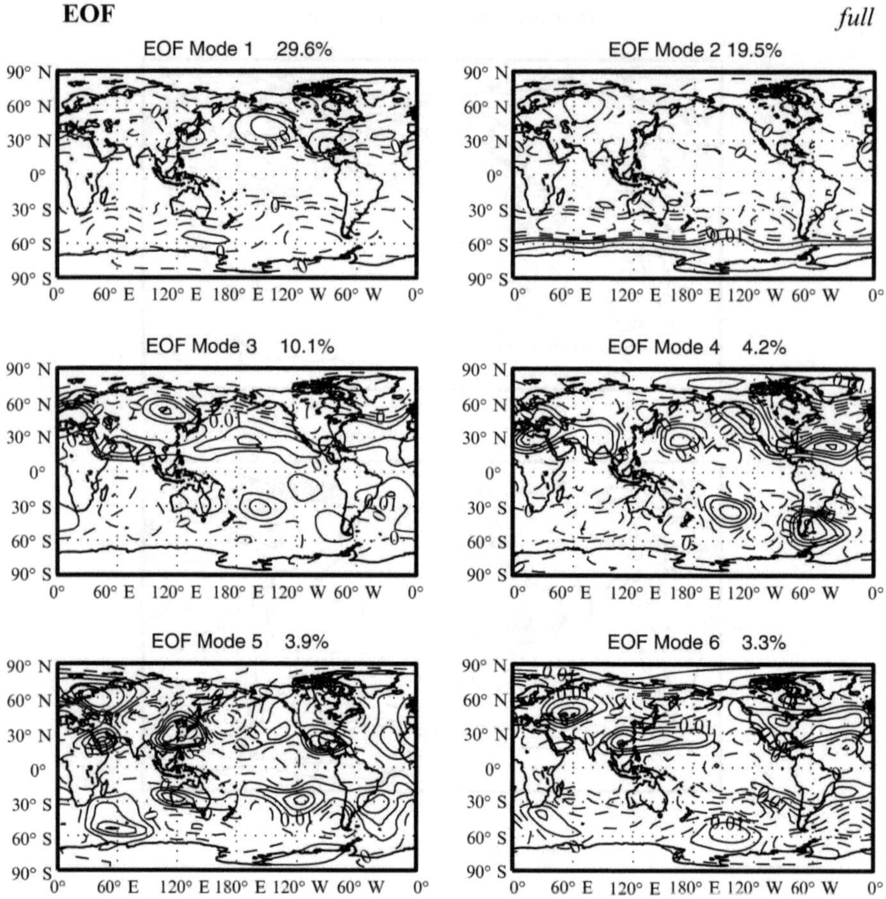

Fig. 4.5 Empirical Orthogonal Functions (EOF) for the Z500 data set. Here are shown the first 6 modes and the amount of variance explained by each mode. The modes have increasingly complex spatial structures, as it is required by the constraint of orthogonality. The higher modes, shown in the following pictures, are increasingly disordered. The physical interpretation of the higher modes is very tricky and it must be done very carefully

data from observations, is always much larger than the length of the time series, that is $m \gg n$. In our case we will therefore obtain a maximum of $n = 34$ EOF. Figure 4.5–4.6 shows the first 12 EOF for the Z500 data. The first three EOF are of course the same as those in Fig. 4.4, but the others have more complicated structures. They present themselves as irregular oscillations in space, with an increasing number of positive and negative centers. This is to be expected as the orthogonality constraint with respect to previous EOF forces them to have zero scalar product. We can also see from the spectrum (Fig. 4.4) that the vast majority of the modes correspond to very small eigenvalues, contributing very little to the total variance. In fact the cumulative variance expressed by the first m modes (bottom of Fig. 4.4) shows that in this case the first 15 modes contribute 80% of the variance, and the rest of the vari-

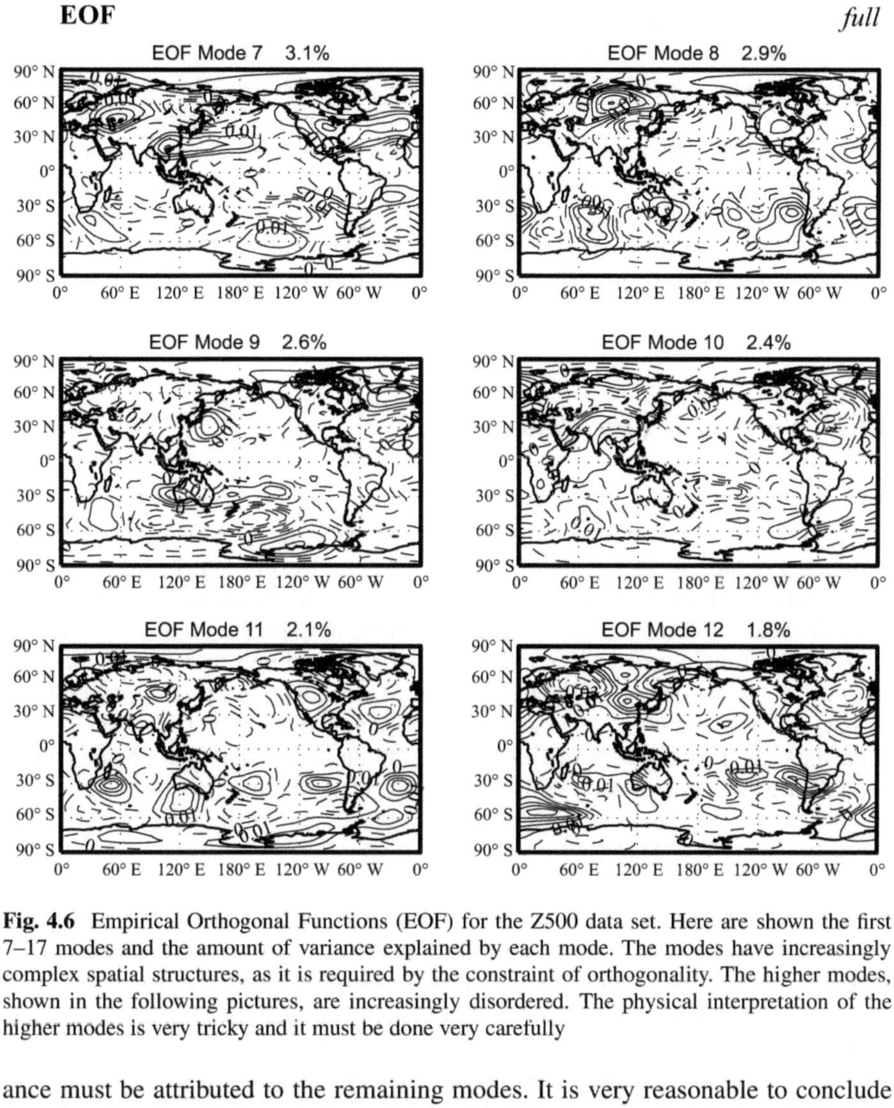

Fig. 4.6 Empirical Orthogonal Functions (EOF) for the Z500 data set. Here are shown the first 7–17 modes and the amount of variance explained by each mode. The modes have increasingly complex spatial structures, as it is required by the constraint of orthogonality. The higher modes, shown in the following pictures, are increasingly disordered. The physical interpretation of the higher modes is very tricky and it must be done very carefully

ance must be attributed to the remaining modes. It is very reasonable to conclude that these latter modes are not important to describe the overall variance of the field, whereas the first modes, corresponding to large fractions of contributed variance, must be of larger relevance.

4.4 Sensitivity of EOF Calculation

We have seen that EOF can be readily calculated, even for large data sets, like the artificial uniformly distributed data with 500 cases of Fig. 4.3. The interpretation of the EOF is provided by the terms of variance explained and recurrent patterns, but if a viable interpretation has to be found, it must rely on a robust determination of the pattern themselves. The SVD algorithm is robust and reliable algorithm, so we are not really concerned with mathematical and/or numerical sensitivities, but with sensitivities deriving from the other possible choices that we can tackle in the definition of the problem itself. A simple mathematical uncertainty is, for instance, that eigenvectors are computed up to a change in sign, as it can be derived from (4.2). Moreover, data can be normalized in different ways. This operation is often done to stress one aspect or another of the data, as one may want to consider a different geographical domain or to analyze a certain area for economy of calculation and space. In the following we will discuss how the EOF react to this kind of changes.

4.4.1 Normalizing the Data

Data can be normalized in several ways. As we have seen when discussing the correlation matrix, the most common normalization is the division by the standard deviation. For the considered multi-variate set this implies dividing by a standard deviation that is different for each station. This approach allows us to compare time series for stations that have large differences in the amplitude of the variability and focus on the time consistency relation among station time-series. As in previous sections, in the following we assume that the vector mean has been removed from the data matrix, so that we can assume that the sample mean is 0. The normalization of the data can be obtained by dividing each column of the data matrix $\mathbf{X}$ in (4.1) by the standard deviation of each station, $\sigma_1, \sigma_2, \ldots, \sigma_m$, that is

$$\mathbf{Y} = \mathbf{D}_\mathbf{X}^{-1}\mathbf{X}, \qquad \text{with} \quad \mathbf{D}_\mathbf{X} = \text{diag}(\sigma_1, \ldots, \sigma_m). \tag{4.5}$$

We can then proceed to compute EOF on the normalized data matrix $\mathbf{Y}$. As we already observed, the covariance matrix of $\mathbf{Y}$ is the correlation matrix of the original data $\mathbf{X}$. However, by normalizing the original matrix, we can compute the EOF at once directly using the SVD of $\mathbf{Y}$, without first computing the correlation matrix.

The EOF of $\mathbf{Y}$ are sometimes called correlation EOF as opposed to the covariance EOF of the unnormalized data that we have seen in the last section. The main differences between the two approaches is that the covariance EOF are going to be biased toward the region of highest standard deviation, so the patterns will try to optimize as much as possible the variation of the field in those regions. On the contrary in the correlation EOF, the normalization equalizes the field variations and so the time series at every station are considered equally important, as a result the patterns will try to described as much as possible the overall spatial variation of the field. The standard deviation has a spatial structure (Fig. 4.7) and the effect of normalization is to reduce the amplitude of the variations in the North Pacific and North Atlantic, whereas the amplitude is expanded in the other regions.

We show in Figs. 4.8 and 4.9 what happens when computing covariance and correlation EOF on our test data set. The main comment is that the first mode is weakly

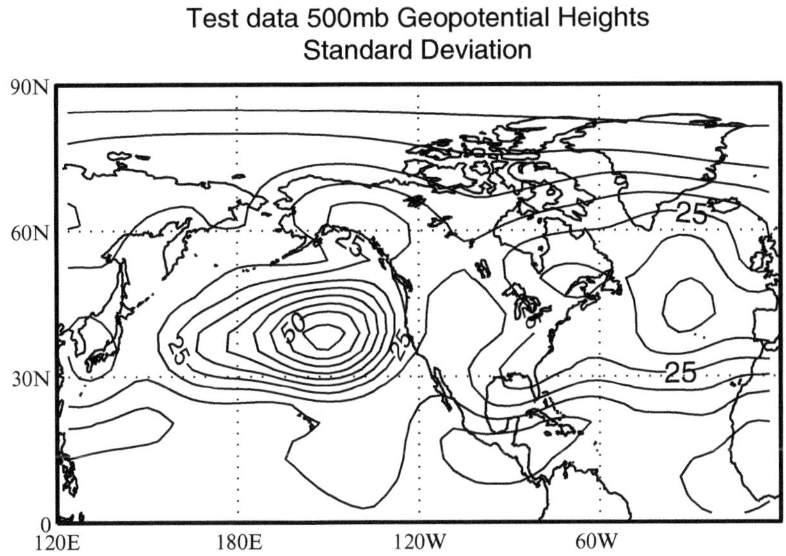

Fig. 4.7 Standard Deviation for the Z500 field

affected, whereas the impact is more noticeable in the higher modes. This is to be expected: the first mode is expressing the major variance mode, so that large, active centers of variation will be well represented with or without normalization. But after the large variations have been removed and higher modes are considered, the impact of normalization increases as the residual variance is different from one case to the other. Here, it is possible to see that correlation and covariance EOF convey different information and we cannot conclude that there is a preferred method regarding normalization. Correlation EOF are to be preferred if the investigator is seeking an overall treatment of all stations, whereas the covariance EOF are simpler to interpret physically, as long as the predominance of the major center of variation is not an impediment to the investigation. In most cases, the most important patterns will remain only slightly affected by the change, as in our example.

The effect of the change in normalization becomes progressively larger as we go up the ladder of modes. This gives some confidence on the reliability of the first mode, but it may cast some doubts on the other modes; how are we going to interpret the other modes?

4.4.2 Domain of Definition of the EOF

To analyze the EOF, in the previous discussion we have selected only a portion of the globe. This is different from the other pictures that represented the entire global data set. The algebraic tools we have used are valid for any row length m that is, there is complete freedom in the choice of the number of stations to perform the EOF anal-

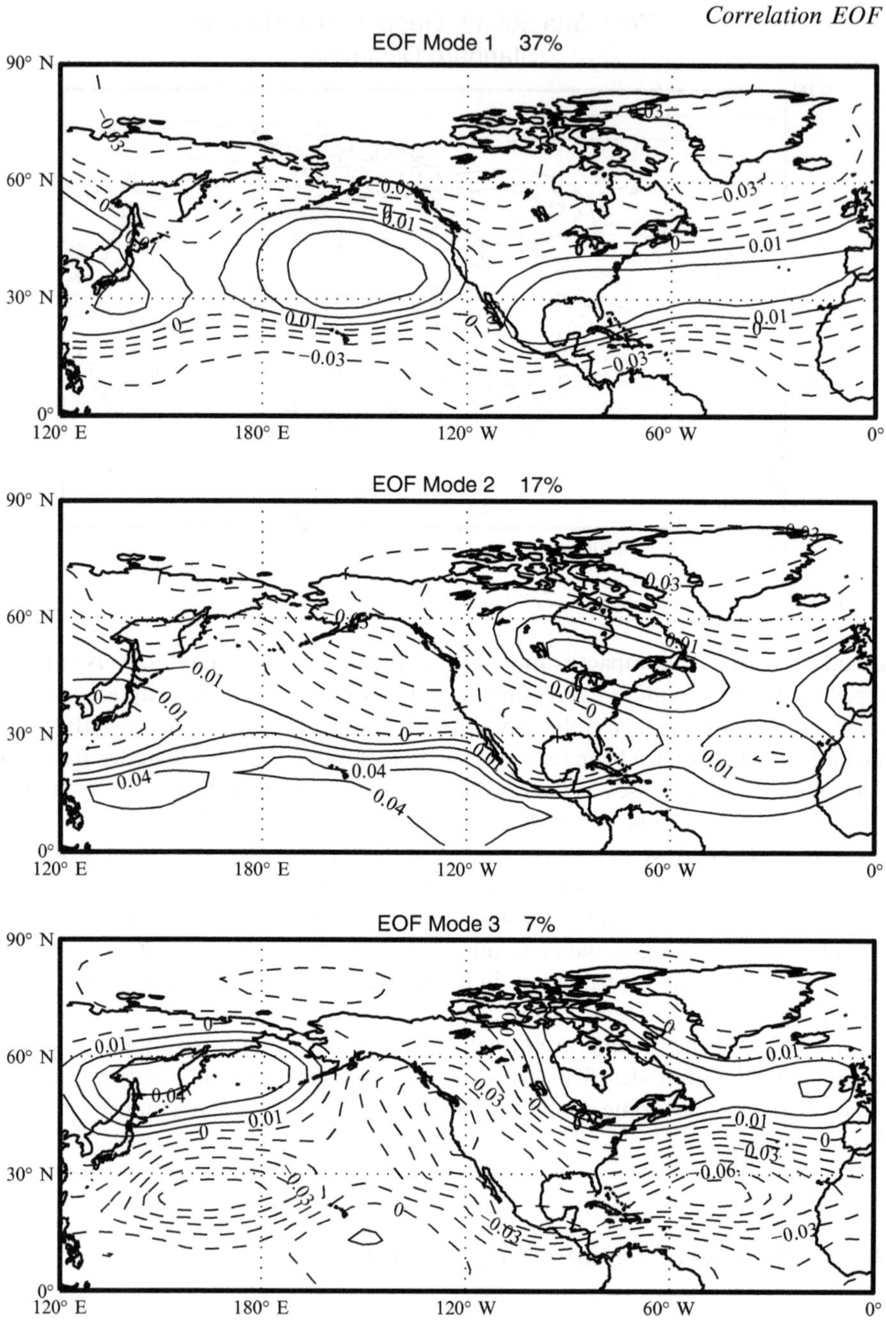

Fig. 4.8 Correlation empirical orthogonal functions (EOF) for the Z500 field

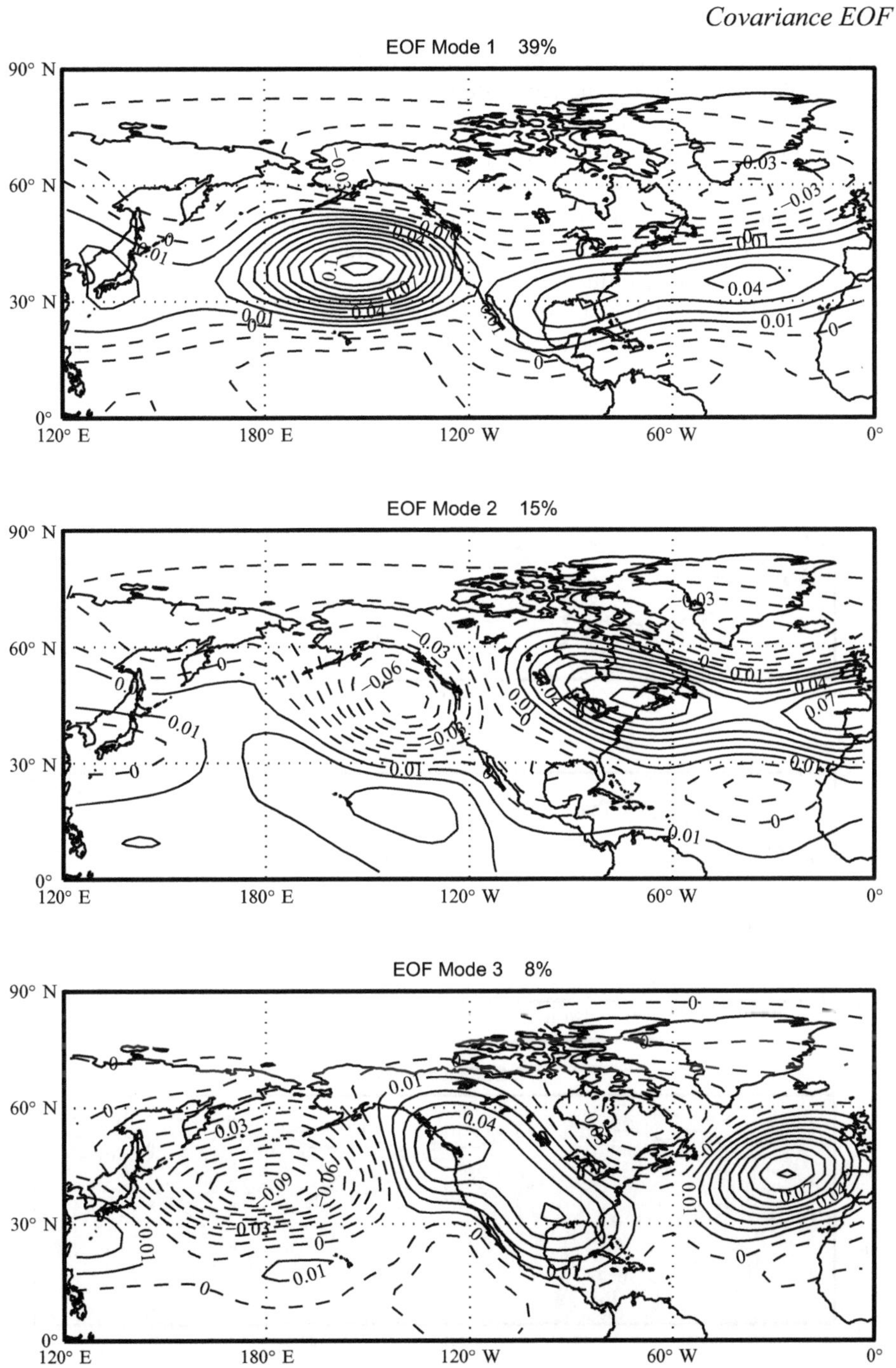

Fig. 4.9 Covariance empirical orthogonal functions (EOF) for the Z500 field

ysis. In fact, in terms of the discussion in Sect. 4.1, a different geographical domain means that a different subset of observation points must be selected. The computation of the EOF follows the same steps for any number of observation points. We will obtain patterns that will try to maximize the variance over the new region, but the optimization of the variance explained is done globally over all stations, so it is possible that eliminating some station may influence the overall optimization and hence generate different patterns. Therefore, it is important to determine whether the obtained patterns are really capturing the major modes of variation we are interested in, without being overly influenced by far stations.

Figure 4.10 shows the result of computing EOF over different geographical domains. The top picture represents the first EOF of the test case for a domain that has

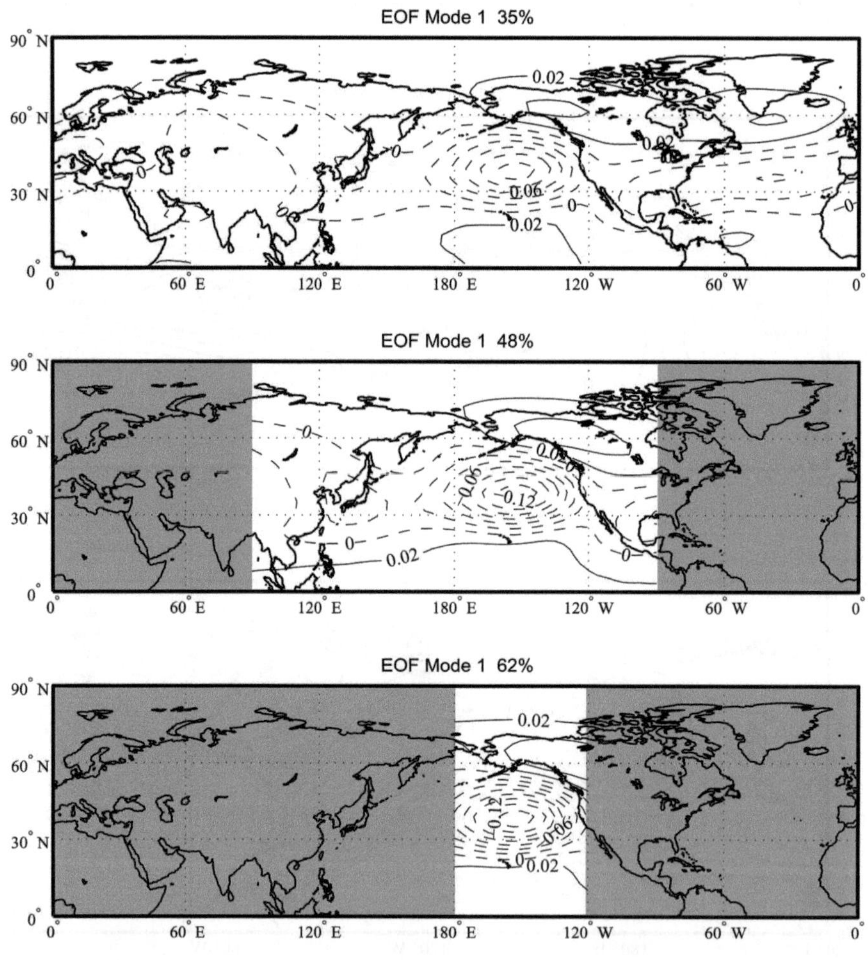

Fig. 4.10 Sensitivity to geographical domain for the Z500 data. The *top picture* represents the first EOF for the Northern Hemisphere, the *other panels* are again the first EOF but excluding the shaded domain

been already altered from the global domain used previously. We have selected here the North Hemisphere. Each panel in the picture shows the same first mode computed on smaller and smaller domains. We can see that the EOF are very consistent from one domain to the next, the computed pattern shape is very similar to each other and the activity centers are correctly identified in each domain. In this case we can be confident that the EOF really represent a major mode of variation. On the other hand, it is interesting to note that the amount of explained variance varies considerably, more than doubling from the hemispheric case to the smallest domain. This reflects the fact that as the domain gets smaller the mode becomes more and more dominant over the total variance in the area and it explains a larger and larger fraction of the variance.

It is important to keep in mind that the smaller modes of variation are dependent on the geographical domain; what we see here is that we do not get spurious influences from areas of low variability on the identification of the areas of high variability. Had we chosen a completely different domain not including the major center in the Pacific, we would have gotten a completely different mode, because that mode would explain most of the variance over the new area, unless we attempt to also explain the variance elsewhere. For instance, the shape over India would have been different had we chosen a domain only over India.

4.4.3 Statistical Reliability

The EOF computed in previous sections are an estimate. They represent the estimate of the true modes of variability performed with the particular available sample. It is very important to give an assessment of the statistical significance of the patterns that have been found with the method. This is a very active and challenging area of research since we are dealing with spatial fields and it is actively investigated, but sometimes a simpler approach can still give a rough idea, that can be used to identify gross problems. A simple idea is to divide the data sets and repeat the analysis, aiming to test the statistical robustness to perturbations in the sampling and distribution of the data. In practice a subset must be defined within the data set and the EOF analysis must be repeated on the smaller sample. The choice of the subset is of course arbitrary: splitting in two the time series, or sub sampling them in some manner are very popular choices. Figure 4.11 shows what happens when the EOF are computed on subsets of the original data consisting of the odd and even years.

The first mode appears rather insensitive to the subsampling, repeating almost identical in the smaller sets and in the total time series. Though this is not a rigorous test, it is usually a reasonably good indication of the absence of major problems in the data sampling.

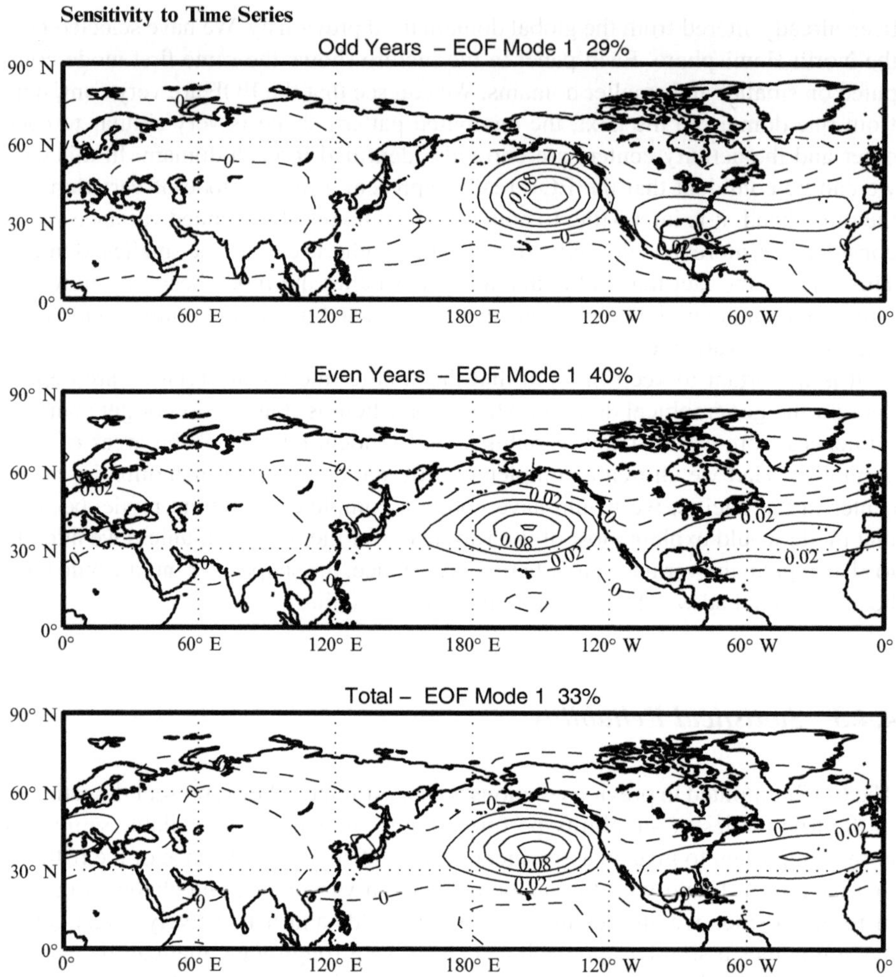

Fig. 4.11 Sensitivity to sampling for the Z500 data. The *top picture* represents the first EOF for the Northern Hemisphere, selecting only the even years of the data set, the *middle panel* is the EOF for the odd years and the *bottom panel* is for the total

Exercises and Problems

1. Show by Matlab computation that the covariance matrix of the (tall rectangular) data matrix

$$\mathbf{X} = \begin{pmatrix} 1 & 1 & 1 \\ 1 & 0 & -2 \\ 1 & 1 & 1 \\ 1 & 0 & -2 \\ 1 & 1 & 1 \end{pmatrix}.$$

can be obtained both by the command $\mathbf{S} = \texttt{cov}(\mathbf{X})$, as well as by means of the SVD of $\mathbf{X}$. Comment on the computed EOF.
On the one hand, we obtain

$$S = \texttt{cov}(X) = \frac{3}{10}\begin{pmatrix} 0 & 0 & 0 \\ 0 & 1 & 3 \\ 0 & 3 & 9 \end{pmatrix}.$$

The mean vector of X is $\bar{\mathbf{x}}^* = [1, 0.6, -0.2]$, so that the zero mean matrix associated with X is

$$\widehat{X} = X - \mathbf{1}\bar{\mathbf{x}}^* = \frac{1}{5}\begin{pmatrix} 0 & 2 & 6 \\ 0 & -3 & -9 \\ 0 & 2 & 6 \\ 0 & -3 & -9 \\ 0 & 2 & 6 \end{pmatrix}.$$

The SVD of $\widehat{X}$, $[U, \Sigma, V] = \texttt{svd}(\widehat{X})$ yields the matrices

$$V = \frac{1}{\sqrt{10}}\begin{pmatrix} 0 & 0 & \sqrt{10} \\ 1 & 3 & 0 \\ 3 & -1 & 0 \end{pmatrix}, \qquad \Sigma = \text{diag}(\sqrt{12}, 0, 0),$$

from which, for $n = 5$, we obtain $\frac{1}{4}V\Sigma^2V^* = S$. We observe that the only significant EOF is given by the scaled version of the vector $\mathbf{v} = [0, 1, 3]^*$, associated with the only nonzero singular value.

2. Do the same for the matrix $\mathbf{X}^*$. *The covariance matrix is*

$$S = \text{cov}(X^*) = \begin{pmatrix} 0 & 0 & 0 & 0 & 0 \\ 0 & \frac{7}{3} & 0 & \frac{7}{3} & 0 \\ 0 & 0 & 0 & 0 & 0 \\ 0 & \frac{7}{3} & 0 & \frac{7}{3} & 0 \\ 0 & 0 & 0 & 0 & 0 \end{pmatrix},$$

while the SVD of $X^* - \mathbf{1}\text{mean}(X^*)$ yields

$$\Sigma = \text{diag}(\sqrt{28/3}, 0, 0, 0, 0). \quad V = \begin{pmatrix} 0 & 1 & 0 & 0 & 0 \\ -\frac{1}{\sqrt{2}} & 0 & 0 & -\frac{1}{\sqrt{2}} & 0 \\ 0 & 0 & 1 & 0 & 0 \\ -\frac{1}{\sqrt{2}} & 0 & 0 & \frac{1}{\sqrt{2}} & 0 \\ 0 & 0 & 0 & 0 & 1 \end{pmatrix},$$

from which the result follows.

4.5 Reconstruction of the Data

The interpretation of the EOF via the SVD has also another important consequence. The SVD decomposition of the data matrix

$$\mathbf{X} = \mathbf{U}\boldsymbol{\Sigma}\mathbf{V}^* \tag{4.6}$$

with $\mathbf{V} = [\mathbf{v}_1, \ldots, \mathbf{v}_n]$, can be written in terms of each column of the matrix as

$$\mathbf{x}_k = \sum_{i=1}^{q} \mathbf{u}_i \sigma_i \mathbf{v}_i(k), \qquad q \leq \min\{m, n\} \tag{4.7}$$

where we can see that the data can be expressed as a linear combination of the $\mathbf{u}$ vectors, weighted by the singular values (the square root of the variance explained) and by the kth component of the vectors $\mathbf{v}$. The summation extends to q vectors depending on the number q of nonzero singular values of $\mathbf{X}$. Equation 4.6 can also be rearranged to provide the interpretation for the vectors $\mathbf{u}$. We have $\mathbf{X} = \mathbf{U}\boldsymbol{\Sigma}\mathbf{V}^*$, so that, multiplying from the left by $\mathbf{U}^*$ and exploiting the orthogonality of its columns, we get $\mathbf{U}^*\mathbf{X} = \boldsymbol{\Sigma}\mathbf{V}^*$. Substituting in (4.7) we obtain for each location vector in $\mathbf{X} = [\mathbf{x}_1, \ldots, \mathbf{x}_n]$,

$$\mathbf{x}_k = \mathbf{U}\mathbf{U}^*\mathbf{X}\mathbf{e}_k = \sum_{i=1}^{q} \mathbf{u}_i \langle \mathbf{u}_i, \mathbf{x}_k \rangle, \qquad k = 1, \ldots, n. \tag{4.8}$$

This equation shows that the data can be reconstructed as a linear combination of the EOF, with coefficients obtained by projecting each data vector onto each EOF. In this way, the singular value decomposition indicates the minimum number of vectors that is needed to describe the data space. Individual EOF can still have no contribution to a certain data map, if the projection of the data vector onto the EOF is tiny.

Exercises and Problems

1. Consider the matrix of the first exercise of Sect. 4.4.3. Show that $\widehat{\mathbf{X}} = \mathbf{X} - \mathbf{1}\bar{\mathbf{x}}^*$ can be fully reconstructed as a rank one matrix.
   Using the SVD of $\widehat{\mathbf{X}}$, we can write

$$\widehat{\mathbf{X}} = \mathbf{u}_1 \sqrt{12} \mathbf{v}_1^*.$$

2. Consider the following matrix:

$$\mathbf{A} = \mathbf{U}\Sigma\mathbf{V}^*$$

$$= \begin{pmatrix} \cos\theta_1 & \sin\theta_1 & 0 \\ -\sin\theta_1 & \cos\theta_1 & 0 \\ 0 & 0 & -1 \end{pmatrix} \mathrm{diag}(1, 10^{-3}, 10^{-6}) \begin{pmatrix} 1 & 0 & 0 \\ 0 & \cos\theta_2 & \sin\theta_2 \\ 0 & -\sin\theta_2 & \cos\theta_2 \end{pmatrix}^*,$$

with $\mathbf{U} = [\mathbf{u}_1, \mathbf{u}_2, \mathbf{u}_3]$, and $\mathbf{V} = [\mathbf{v}_1, \mathbf{v}_2, \mathbf{v}_3]$ and $\Sigma = \mathrm{diag}(\sigma_1, \sigma_2, \sigma_3)$, $\theta_1 = \pi/6$ and $\theta_2 = \pi/8$. Numerically show that

$$\|\mathbf{A} - \mathbf{u}_1\sigma_1\mathbf{v}_1^*\|_2 = \sigma_2, \qquad \|\mathbf{A} - \mathbf{u}_1\sigma_1\mathbf{v}_1^*\|_F = \sqrt{\sigma_2^2 + \sigma_3^2}.$$

This result is very general, and provides the error (in the given norm) occurring in the reconstruction of the given matrix **A** by means of the first few terms in the SVD.

4.5.1 The Singular Value Distribution and Noise

The set of singular values is sometimes called the spectrum, in analogy with the spectrum of eigenvalues. We have seen examples of spectra in our constructed case to show the role of carefully selected data with known composition of the basis vectors and with known statistics of the data distribution. Reality is much more complex. Multiple time scales and statistics are present in the data and the problem is to identify in the total variability the sector of insightful variability where we think the real physical processes may be at work. The issue is further complicated because there is no clear and absolute distinction between *signal* and *noise*, as the separation is problem dependent. What is noise for some investigators may be signal for others, or even for the same investigator at a different time.

The possibility of expanding the data on the EOF that we have seen in the preceding section gives us a very nice opportunity. The results we have obtained are purely algebraic with no reference to statistics, but the issue of statistics comes in when we consider again Fig. 4.12 (the spectra for real Z500). In this case, the spectrum is similar to the preceding examples, there is a small number of large singular values, with a long tail of rapidly decreasing values, representing modes that contribute less and less to the variance. As with all numerical calculations, we have to check that these values are really non-zero by checking them against the *numerical zero* of the calculation. A very useful test for this purpose is to assume that a singular value of a matrix **A** is nonzero if

$$\sigma > \epsilon \, \|\mathbf{A}\|_2 \,, \tag{4.9}$$

where ϵ is the smallest number represented in the floating point arithmetic in use (in MATLAB double precision computation this is 2.2204e-16) and $\|\mathbf{A}\|_2$ is the 2-norm of **A** (cf. Sect. 2.8), namely its largest singular value. In other words, (4.9) says that the numerical zeros are those values that, when normalized by the largest singular value, are smaller than the smallest number represented in the employed arithmetic. A quick inspection reveals that all the sigma's in Fig. 4.12 pass the test, when the

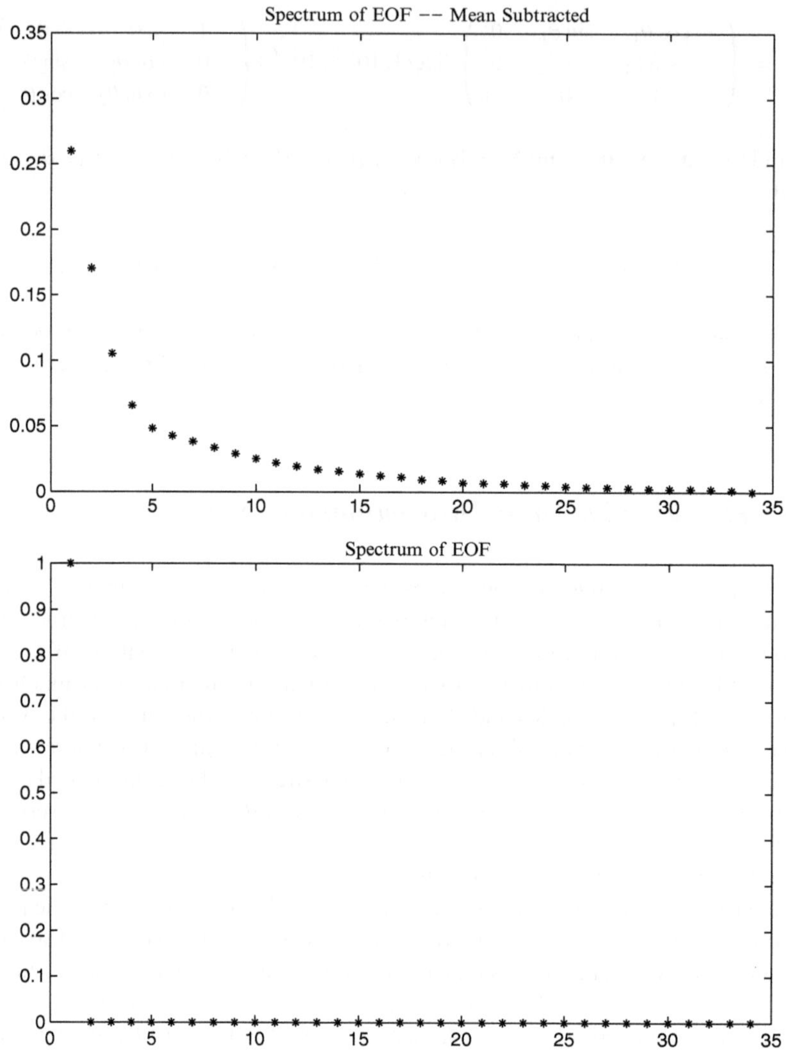

Fig. 4.12 Singular value distribution for a real case (global Z500 data sets). *Top panel*: the mean value of the data is subtracted before calculation of the EOF. *Bottom panel*: the mean is not removed. Values are normalized with the trace to express explained variance

double precision computation in MATLAB is taken into account, actually they pass the test very well, with a difference of several orders of magnitude. We have then eliminated the idea that some of the small sigma's can in fact be zero, that only the finite nature of the computation is preventing from showing up. It is also interesting to note that in the case we are mostly interested in here, that is for $m \geq n$, there is exactly one zero singular value when the mean is removed, because the subtraction of the mean reduces the degrees of freedom of the data by one.[2]

The conclusion that we need to retain all modes may in fact be premature. There is another issue that we need to investigate. We have computed the EOF on the available data sample, but it is not at all clear how representative of the *true* EOF they are. In practice, we have to realize that our computation is only an estimate of the EOF of the population from which we have extracted a sample, and we need to investigate how accurate this estimate is. In particular, we need to evaluate the probability that some of the sigma's are zero just because of the choice of the members of the sample. Some sigma's can really be zero and correspond to degrees of freedom that do not contribute to the variability of the field, but others may appear nonzero just because of our particular sampling. An EOF analysis is therefore incomplete without some consideration on the robustness of the results and their sensitivities to changes in the sampling or in other aspects of the procedure.

The EOF have identified some patterns corresponding to observation points that vary together in an organized manner, but each observation point may have variance that is uncorrelated from other points, from the point of view of the spatial analysis of variance that is considered noise.

Mathematically the EOF will tend to fit also those components, thus generating a fictitious pattern. This is one of the reasons that explains why the higher order EOF have very complex patterns. They try to fit the variance point by point: a desperate job since it is mostly uncorrelated. This portion of variance is not really interesting, but we can exploit this property of the EOF, because we can then use it to generate data that is free of the noise component, simply by reconstructing the data sets retaining only the higher modes corresponding to covarying modes (cf. Sect. 4.5).

Another example is shown in the following pictures. A two-dimensional wave is propagating in a square domain from left to right. The wave is a fairly regular sine wave, but a substantial amount of noise is superposed. At any time the wave pattern is substantially distorted by the noise (Fig. 4.13). In Fig. 4.14 we display the EOF of the time evolution, obtained by considering as observation points the local position at which the wave is observed to pass.

The EOF recover fairly quickly the coherent pattern of the propagating wave and the first two modes explain most of the total variance. We can also see how propagation is represented by EOF usually employing two modes that are in quadrature and fairly similar in distribution. This indicates that those modes are two phases of the same propagating pattern. The noise is relegated to higher modes; having added a significant amount of noise, these modes are not insignificant. The totally

---

[2] More precisely, using $\bar{\mathbf{x}} = \frac{1}{n}\mathbf{X}\mathbf{1}$, we have

$$\mathbf{X} - \bar{\mathbf{x}}\mathbf{1}^* = \mathbf{X}(\mathbf{I}_n - \frac{1}{n}\mathbf{1}\mathbf{1}^*).$$

Since the matrix $\mathbf{I}_n - \frac{1}{n}\mathbf{1}\mathbf{1}^*$ has rank $n-1$, the relation above shows that $(\mathbf{X} - \bar{\mathbf{x}}\mathbf{1}^*)$ has rank not greater than $\min\{n-1, m\}$. Therefore, the scaled covariance matrix $(\mathbf{X} - \bar{\mathbf{x}}\mathbf{1}^*)(\mathbf{X} - \bar{\mathbf{x}}\mathbf{1}^*)^*$ has rank at most $n-1$, if $m \geq n$.

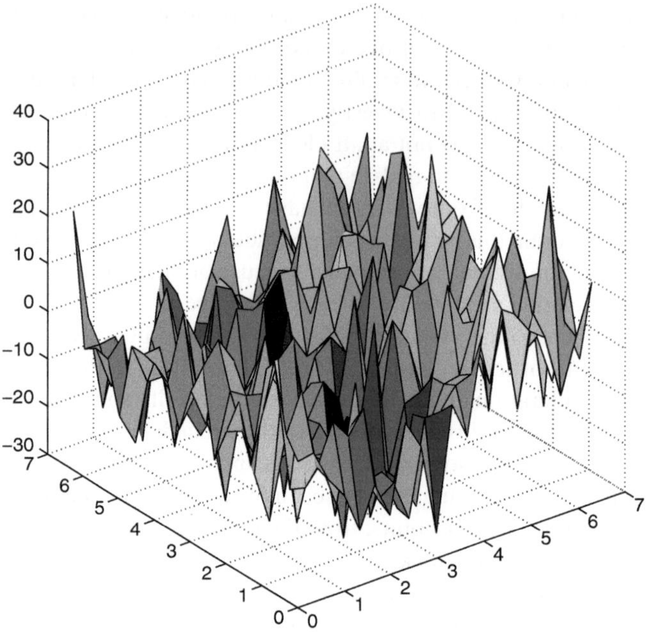

Fig. 4.13 The propagating wave at an arbitrary instant in the propagation. The average is substantially distorted by the superposed noise

incoherent distribution of these patterns is a very clear indication that in this case the EOF are trying to fit the individual noise at each observation point.

4.5.2 Stopping Criterion

The division between signal and noise is somewhat arbitrary: how can we decide when to stop? Though a number of rules to select significant EOF have been proposed, their statistical foundation is tenuous. It is better to realize that in practice the choice is essentially driven by empirical considerations. In the atmospheric literature, North (North et al. 1982) has proposed the following *rule of thumb*. He estimated *typical errors* using the eigenvalues of the covariance matrix, λ_k, and the number of statistically independent samples in the data

$$\Delta\lambda_k \approx \sqrt{\frac{2}{n}}\lambda_k. \tag{4.10}$$

The rule can then be stated by saying that when the error is larger than or comparable to the spacing between neighboring eigenvalues, then the sampling error on the EOF will be comparable to the size of the neighboring EOF. This rule of thumb is often

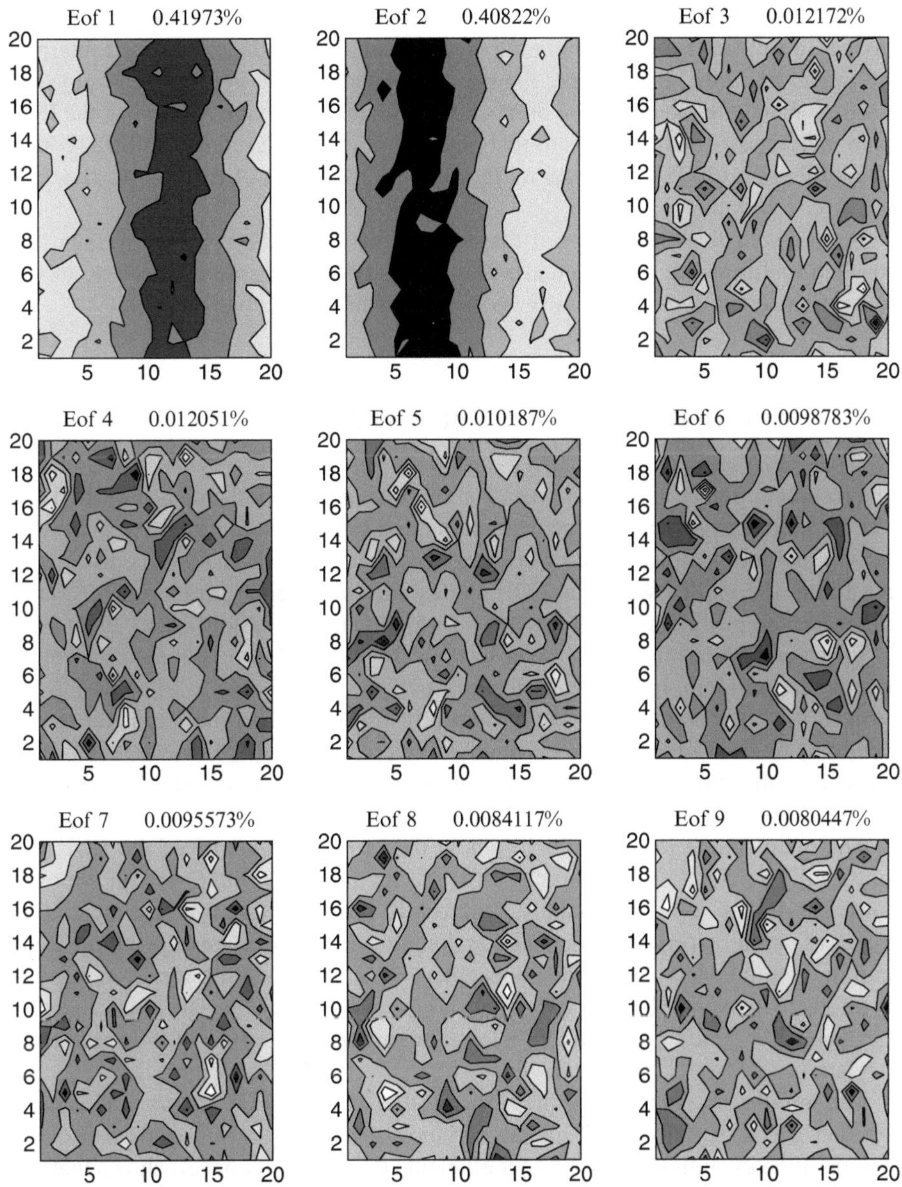

Fig. 4.14 Empirical orthogonal functions for a propagating wave. The wave propagates from *left* to *right* in the picture. The first two modes correspond to the waves, the other modes apparently try to capture the random noise that had been added to the wave

consistent with another highly employed empirical rule that basically looks for sharp changes in the convergence to zero of the eigenvalues, the so-called "elbow". Figure 4.12 shows a sharp change in the curve slope around the fifth and sixth eigenvalues.

The larger eigenvalues are of similar magnitude and they would not pass North's rule anyway. The idea is then to retain the eigenvectors before the change and interpret the others as *noise*. In practice, there is no objective rule and only the physical discussion and the identification of mechanisms can support the patterns of the EOF from merely statistical patterns to real physical objects. We refer to Jolliffe (2002), Quadrelli et al. (2005) for a general thorough presentation of truncation strategies and error estimation.

From a statistical viewpoint, if the data follow a normal distribution, then it is possible to formulate an hypothesis test as a stopping strategy. However, in our context, the condition of normality is often too restrictive to be feasible.

4.6 A Note on the Interpretation of EOF

The technical preparation of the data for analysis and the performance of the EOF computation themselves may be demanding and time-consuming, to the point that the original motivations for the work tend to fade. The EOF analysis is just a tool, and filling the gap between the EOF modes and the original problem is of major importance to be able to deduce useful information on the system under analysis.

Interpretation of the EOF is not a form of divination, but it is about connecting them to the problem. In physical sciences, such an interpretation is often cast in the form of an "identification" of the system physical modes. In more general terms, EOF may be used to identify recurrent patterns of variations. However, the mathematical nature of EOF generates common mistakes that lead to misinterpretation of the obtained results.

The main source of the problem lies in the orthogonal nature of the EOF. Orthogonality forces a special structure for the modes and sometimes can alter the dominant modes quite significantly. Figure 4.15 shows an idealized example. Two modes of variations are set in rectangular domain without overlapping with the same amplitude (top panels) as two distinct centers of action. The time variation is given by another sine wave (bottom left panel). These waves have the same phase so that they show the same time variation. Each of them represents 50% of the variance in the data.

The middle panel shows the first three EOFs. The first EOF shows both centers together and it explains 100% of the variance, the higher EOF are just noise and also the time evolution of the EOF coefficients (right bottom panel) shows that the first EOF captures the correct time behavior. This example shows very clearly how efficiently the EOF will capture any co-varying phenomenon, when the center of actions are separated, but of course does not provide any indication of the causes of the variations. The EOF aim at maximizing the variance with the smallest possible number of modes, and this can be done very nicely with only one mode in this case, but interpreting the first EOF mode as the only mode of the system is incorrect because by construction we had two.

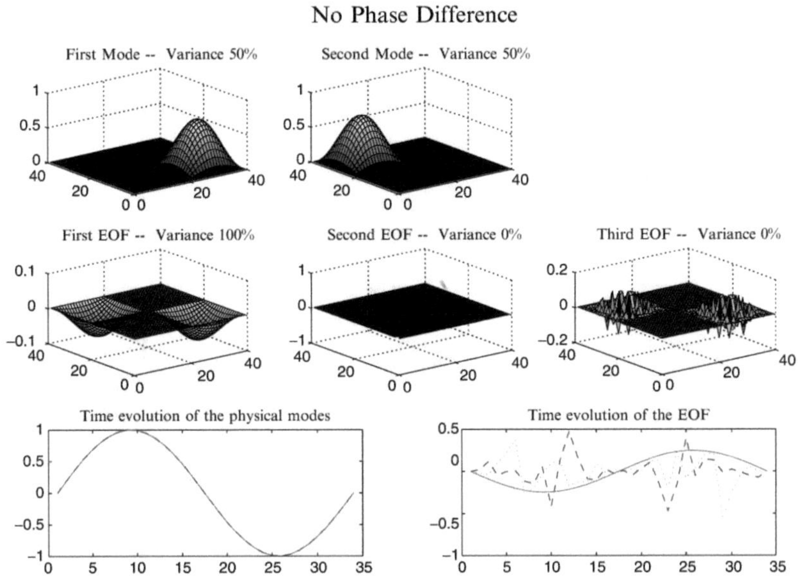

Fig. 4.15 An idealized example. Two modes are considered within a rectangular domain with a simple sine shape with the same amplitude (*top panels*). The first three EOF are shown in the *middle panel*. The first mode explains 100% of the variance and the remaining modes represent just noise. The time evolution of the modes is shown in the *left bottom panel*. The modes are in phase so that the time plots are exactly on top of each other. The EOF coefficients in time (*right bottom panel*) show the correct behavior for the first mode, but just noise for the higher modes

By varying the phase relation in time between the modes the situation changes. In Fig. 4.16 we have used the same data as in Fig. 4.15, but we have changed the time phase in a way that the two modes are uncorrelated (bottom left panel). The variance is split evenly between the modes, except for sampling errors. The EOF now identifies correctly the existence of two distinct modes; moreover, the estimation of the explained variance is in the right ball park. The higher modes are noise and they count for a negligible fraction of the variance anyway.

The results for the data in quadrature in time suggest that if the modes are uncorrelated then the EOF can pick them up rather easily. We tried that in Fig. 4.17 where we assigned as a time evolution random data normally distributed with mean zero and standard deviation one. In this case the EOF generate two modes that are quite different from the "modes". The modes have two centers of action and the first one has the same sign everywhere in the domain; the second must have opposite signs, so that the overlapping space integral between the two modes is zero, as required by the orthogonality condition for the EOF. This example is highly influenced by the sampling error. The 34 time samples used in this case are simply not sufficient to correctly identify the independence of the two modes, but it is easy to check that by increasing the number of time samples the modes are recovered correctly.

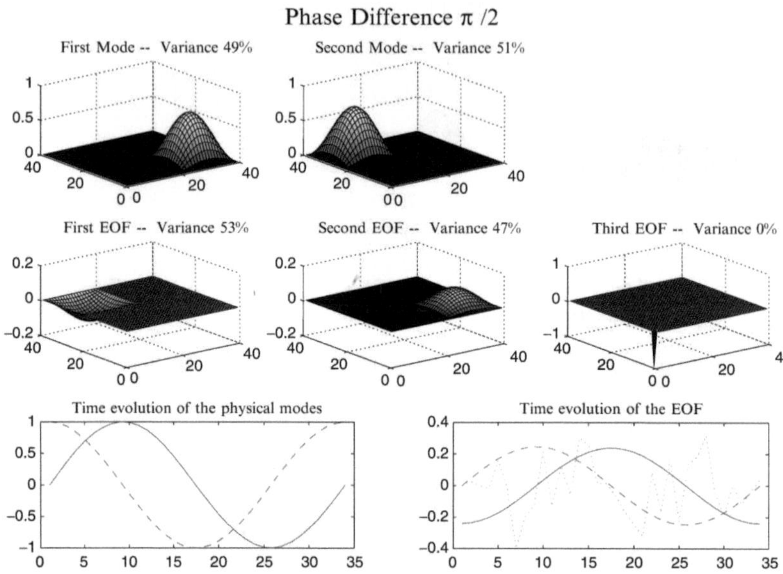

Fig. 4.16 As in Fig. 4.15 but for the case when the two modes are in quadrature. The time evolution (*bottom*) shows that they are in quadrature. In this case the EOF modes capture two distinct modes. The estimation of the variance explained is also good and it will get better as the statistics is improved

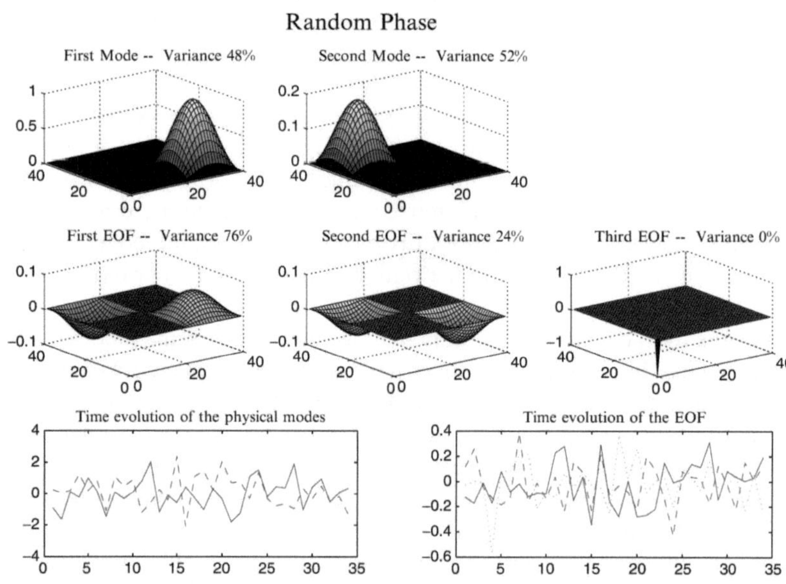

Fig. 4.17 As in Fig. 4.15 but for the case when the two modes are uncorrelated in time. The time evolution (*bottom*) coefficients are random numbers extracted from a normal distribution with zero mean and unit standard deviation. In this case the EOF modes capture two distinct modes

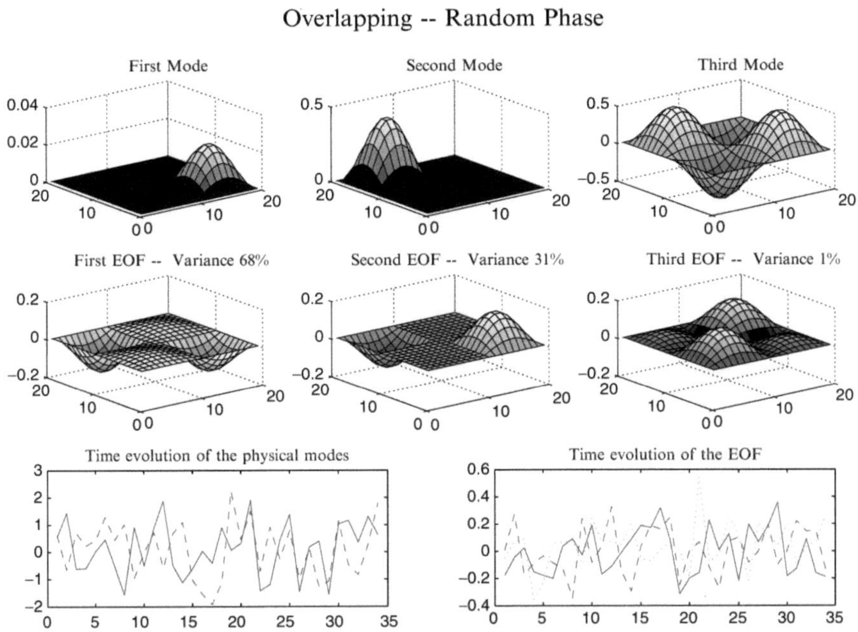

Fig. 4.18 As in Fig. 4.15 but for the case when there are three modes uncorrelated in time. The time evolution (*bottom*) coefficients are random numbers extracted from a normal distribution with zero mean and unit standard deviation. In this case the EOF modes capture two distinct modes

The general picture changes completely in case the analyzed *physical* modes are not orthogonal. In Fig. 4.18 we display a situation when we have three modes: the first two modes are the same as the previous examples, but we have added a third mode that is not localized to only one quarter of the domain but it extends over the entire domain, overlapping the other modes. The EOF are shown in the middle panel and it is clear that they have significant deviations from the target modes. Of course sampling errors are still playing a role, but the main problem here is that the overlap makes it really hard for the orthogonal EOF to identify the modes.

These examples show that the interpretation of the EOF is often a delicate issue and it is important to keep in mind that the orthogonality constraint will very easily generate wave-like patterns, which may be easily misinterpreted as oscillations.

Exercises and Problems

1. Show by Matlab computation how the EOF modes change in Fig. 4.17 if the sample size is increased.
2. Show by Matlab computation how the EOF modes change in Figs. 4.15–4.17 if the phase relation between the modes is modified.

Chapter 5
Generalizations: Rotated, Complex, Extended and Combined EOF

5.1 Introduction

We have seen in the last section that the difficulty in identifying real physical patterns from EOF stems from their orthogonal nature. Orthogonality translates into the fact that typical patterns appear in secondary (higher order) EOF. Very often the first EOF has little structure, the second has a positive and a negative center, the third more centers and so on, in a way so as to maintain orthogonality.

In fact, the overall structure of the EOF is often determined by the geometrical shape of the domain chosen and different data on the same domain, with different covariance relations, may actually result in similar EOF. Moreover, in the preceding chapter we have seen the issue of sensitivity to partitioning the analysis domain into subdomains and we can interpret it as a case of the overall sensitivity to the domain shape. Another point is that EOF are obtained by trying to maximize the amount of total variance explained by a single mode. It is possible that the resulting optimized modes are difficult to interpret physically, either because the real relation is localized and the EOF are spreading it, creating artificial nonlocal relations, or because the EOF are so close in terms of eigenvalue separation, that the numerical techniques cannot really distinguish between them.

The risk of creating misleading or illusory representations of relations within the data is particularly troubling. The most common situation in which these malfunctioning can arise is when data represent localized variances. In this case the EOF will try to fit globally the domain under consideration, with as few modes as possible, generating first EOF (low order modes) with very large structures. For instance, it is possible that the first mode indicates anomalies all of the same sign, whereas the data do not indicate that there is ever a time when all stations were reporting such a one-sign pattern. This may be a fiction created by the EOF trying to maximize the variance explained. This is another way of realizing that the EOF have no way to guess the physical relations within the data.

A. Navarra and V. Simoncini, *A Guide to Empirical Orthogonal Functions for Climate Data Analysis*, DOI 10.1007/978-90-481-3702-2_5,

5.2 Rotated EOF

The problems mentioned in the previous section do not mean that the EOFs must be ditched. They simply indicate that much care must be taken in producing and interpreting EOF and it is not simply a matter of using some canned routines out of a package.

Unfortunately we cannot produce a recipe for the intelligent use of EOF, but some techniques have been devised to make a misuse of EOF less probable. Rotation is one of them.

Simply put, we need to transform the EOF to another system of coordinates, exploiting the freedom to choose a different basis in the data space. We have seen from (4.3) that the EOF can be seen as the eigenfunctions of the covariance or correlation matrix, that is

$$\mathbf{S} = \mathbf{U}\boldsymbol{\Sigma}^2\mathbf{U}^*. \tag{5.1}$$

The basic indeterminacy in the EOF can be seen if we consider a similarity transformation that changes the matrix $\mathbf{S}$ into a different matrix. More precisely, let $\mathbf{T}$ be an $n \times n$ nonsingular matrix. Then we can write

$$\mathbf{G} = \mathbf{T}\mathbf{S}\mathbf{T}^{-1}, \quad \text{that is} \quad \mathbf{S} = \mathbf{T}^{-1}\mathbf{G}\mathbf{T}.$$

Substituting in (5.1) we obtain $\mathbf{T}^{-1}\mathbf{G}\mathbf{T} = \mathbf{U}\boldsymbol{\Sigma}^2\mathbf{U}^*$, and the eigenvalue decomposition of the non-Hermitian matrix $\mathbf{G}$ is

$$\mathbf{G} = (\mathbf{T}\mathbf{U})\boldsymbol{\Sigma}^2(\mathbf{U}^*\mathbf{T}^{-1}). \tag{5.2}$$

The similarity of $\mathbf{G}$ and $\boldsymbol{\Sigma}^2$ (cf. Sect. 2.7) implies that the transformed matrix $\mathbf{G}$ has the same eigenvalues as $\mathbf{S}$. However, the eigenfunctions have been transformed according to

$$\mathbf{U_T} = \mathbf{T}\mathbf{U}. \tag{5.3}$$

Owing to the non-orthogonality of $\mathbf{T}$, the columns of the new matrix $\mathbf{U_T}$ are correlated. In the derivation above, the aim is to exploit the freedom of choosing $\mathbf{T}$ so as to add some constraints to the EOF. The purpose is to alleviate some of the problems discussed earlier, especially the problem of the generation of patterns that cannot be reconciled with the expected physical relations in the data.

Whenever $\mathbf{T}$ is non-orthogonal, the transformed EOF represent an oblique coordinate system for the given data. Therefore, in this case, they are called *oblique* EOF, and each of them has a nonzero projection (correlation) on each other. On the other hand, if $\mathbf{T}$ is an orthogonal matrix, we can refer to it as a rotation matrix, $\mathbf{Q}$, and we write

$$\mathbf{U_Q} = \mathbf{Q}\mathbf{U}.$$

The choice of rotation is arbitrary, although we may want to require that some "nice" pattern feature be emphasized after rotation. In the most popular case, we

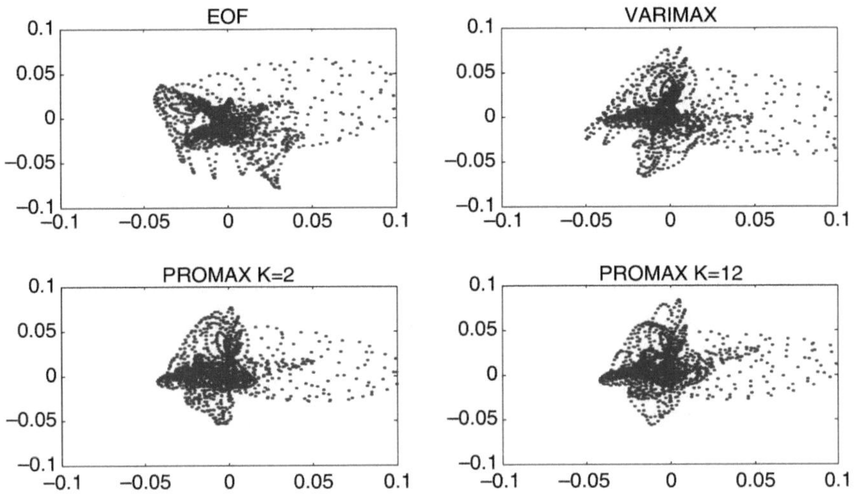

Fig. 5.1 Simple structure. The plots are scatterplot of the value of the EOF for each spatial points by pairs of modes, in this case EOF1 and EOF2. The standard EOF yield a situation in which there are several points where both modes have nonzero values (*top left panel, bottom right quadrant*). Rotated EOFs reduce such an effect, aligning the point values along the axis (*top right panel*), the maximum effect is obtained by relaxing the constraint of orthogonality with the oblique modes of PROMAX (*bottom panels*)

require that the spatial variance be concentrated in as few points as possible, to obtain *simpler* patterns. The definition of what simple is, is far from being straightforward. The issue has been discussed at length in the specialized literature, but a rigorous definition of "simple" is elusive. A set of empirical principles to describe the properties of simple structures has been proposed, but the overall philosophy of the existing school of thought essentially reduces to trying to concentrate the coefficients of the EOF in few modes, in such a way that for each variable, i.e. for each spatial point in our examples, only a very small number of EOF is needed to explain the variance. This is easily checked by producing scatter plots of the EOF modes two at a time. Figure 5.1 shows an example of such a plot, shown here for the case of the marine SST in our test dataset. The EOF are not well separated in the sense that there are several points for which both EOF1 and EOF2 have nonzero elements, as it can be seen by the alignment of the points along the diagonal of the third quadrant in the top left panel of Fig. 5.1. These points are positions where both EOF1 and EOF2 are needed to describe the variance in that point. Ideally, we would like to separate as much as possible the variance in such a way that distinct EOF describe most of the variance at separate points. This means that different EOF have to reach large values in different places, so that if EOF1 has large values in certain points then the higher modes must have small values in those same points. The top left panel of Fig. 5.1 shows that the EOF do not guarantee this property: there is a whole class of points where both EOF1 and EOF2 have important amplitudes.

Orthogonal Rotations

We can exploit the arbitrariness in the coordinate system definition to try to transform the original EOF to another coordinate system that yields a better separation of the EOF for each spatial points, i.e. simple structure. In principle the transformation is quite unrestricted, but orthogonality is a desirable quantity if we want to separate the variance of the field under examination.

The following panel (top right) shows a popular method to introduce some simple structure and maintain orthogonality. This result can be achieved by requiring that **Q** is such that the new pattern minimizes a functional, sometimes called the *simplicity functional*, that provides a distribution measure of spatial variance. The choice of the functional is very important and there are no rules to prescribe it, but a very popular choice is the VARIMAX method, where the functionals are chosen so that the rotated patterns maximize the functional

$$F_R(u_1, u_2, \ldots, u_r) = \sum_{k=1}^{r} f(u_k), \text{ with } \quad f(u_k) = \frac{1}{n}\sum_{i=1}^{n} p_{ik}^4 - \frac{1}{n^2}\left(\sum_{i=1}^{n}(p_{ik}^2)\right)^2,$$

or with

$$f(u_k) = n\sum_{i=1}^{n}(\frac{p_{ik}}{h_i})^4 - \left(\sum_{i=1}^{n}(\frac{p_{ik}}{h_i})^2\right)^2,$$

where the p_{ik} are the grid-point values of the kth EOF patterns (u_k) that we are trying to rotate, the h_i are the point-by-point standard deviations (communalities) of the r patterns we are rotating; see, e.g., Harman (1976). Both forms aim at maximizing the spatial variance of the EOF modes by concentrating the point values toward zero or one: the first example is known as the *raw VARIMAX* and the second one as the *normal VARIMAX*. Below is a possible simple Matlab implementation of the normal VARIMAX procedure, closely following Harman (1976).

```
function [coef,u,varrotated]=eofrot(z,ind,index)
%
% Algorithm for eof rotation with the varimax method
% Inputs:
%         z                    Data Matrix
%         ind                  Index for domain
%         index                indeces of EOFs to be rotated
% Outputs:
%         coef                 Cofficient for the rotated EOFs
%         u                    Rotated Eof in ascending order
%         varrotated           Variance explained by rotated EOFs

[npoints,ntime]=size(z);            %  Time and space points
[uneof,ss,vneof]=svd(z,0);          %  Unrotated EOF for variance
totvar = sum(diag(ss.^2));          %  calculation
```

```
lding = uneof(:,index);
sl    = diag(ss(index,index).^2);
varexpl = sum(sl)/totvar;          % Relative variance explained
                                   % by the unrotated modes
[n,nf]=size(lding);
b=lding;
hl=lding*(diag(sl));

hjsq=diag(hl*hl');
hj=sqrt(hjsq);                   %  Normalize by the communalities
bh=lding./(hj*ones(1,nf));

Vtemp=n*sum(sum(bh.^4))-sum(sum(bh.^2).^2); % VARIMAX functional
                                            % to be minimized
V0=Vtemp;
for it=1:10;                       % Number of iterations
  for i=1:nf-1;               %  Program cycles through 2 factors
    for j=i+1:nf;
      xj=lding(:,i)./hj;           % notation here closely
      yj=lding(:,j)./hj;           % follows harman
      uj=xj.*xj-yj.*yj;
      vj=2*xj.*yj;
      A=sum(uj);   B=sum(vj); C=uj'*uj-vj'*vj;   D=2*uj'*vj;
      num=D-2*A*B/n;             den=C-(A^2-B^2)/n;
      tan4p=num/den;     phi=atan2(num,den)/4; angle=phi*180/pi;
      if abs(phi)> eps;
       Xj=cos(phi)*xj+sin(phi)*yj; Yj=-sin(phi)*xj+cos(phi)*yj;
       bj1=Xj.*hj;                 bj2=Yj.*hj;
       b(:,i)=bj1;                 b(:,j)=bj2;
       lding(:,i)=b(:,i);          lding(:,j)=b(:,j);
      end
    end
  end;
  lding=b;           bh=lding./(hj*ones(1,nf));
  Vtemp=n*sum(sum(bh.^4))-sum(sum(bh.^2).^2);% Update functional
  V=Vtemp;
  if abs(V-V0)<.0001;break;else V0=V;end;
end;

for i = 1:nf             % Reflect vectors with negative sums
    if sum(lding(:,i)) < 0
        lding(:,i) = -lding(:,i);
    end
end
Arot=lding ;                       % rotated eof
coef=z'*Arot(:,1:nf);              % time series for rotated eof

for i=1:nf
   varex(i) = sum(var(coef(:,i)*Arot(:,i)')*(ntime-1));
end

varexplrot = sum(varex)/totvar;  zvar=sum(var(z')*(ntime-1));
[varex,I]=sort(varex);    % Sort in decreasing order of variance
Arot=Arot(:,I);   Arot = fliplr(Arot); varex = flipud(varex');
```

```
varunrotated = sl/totvar;      varrotated   = varex/totvar;
u=zeros([96*48 nf]);
u(ind,1:nf) = Arot(:,1:nf); % Rotated EOF in mapping formats

end
```

The previous function performs a rotation using orthogonal rotations; the sister function eofpromax uses oblique rotations to maximize the spatial variance and can be found in the book Website. The Matlab Statistics Toolbox ((matlab7)) includes a few routines to compute the EOFs (Principal components) and the rotated factors; see Exercise 1 below.

The pictures in Figs. 5.2 and 5.3 show the difference between unrotated and rotated EOF, in this case after a normal VARIMAX rotation has been used. The rotation has been applied to the first ten modes. We can see how the rotation tends to separate the original EOF in a spatial sense. The first unrotated mode (top panel, Fig. 5.2), for instance, is composed of centers of activity, i.e. relative maxima and minima for the patterns of the EOF, that are distributed across the North American continent and the North Atlantic extending well into the European continent.

The rotated equivalent (top panel, Fig. 5.3) shows the emergence of a pattern that is more confined to the North American sector, with small or no amplitude elsewhere. The variation over Europe and Asia is picked up by the higher modes, represented here by modes 3 and 10, that instead tend to accumulate amplitude over the regions where there is little or no amplitude for mode 1. The separation is not perfect, as it can be noticed that mode 3 still has some amplitude in the central Pacific, in correspondence of the centers of mode 1. The effect is larger on the higher modes, and the rotated mode 10 is now more concentrated over Asia, showing a clear pattern from India to the Mediterranean. It is not possible to give a general rule on when rotation is necessary. It is found that when the EOF modes are very close together, i.e. the separation in the eigenvalues is not great, then rotation can disentangle the modes in the previous case between the Pacific and Atlantic modes.

The rotated modes can still be used to decompose the variance, in the sense that each of them explains a certain portion of the variance that can be attributed only to that mode, since the rotated EOF are still mutually orthogonal. The rotated EOF can then be ranked in order of percentage of explained variance.

The issue of rotation is still not widely accepted. Some investigators think that rotation should become the standard and therefore recommend to rotate all modes before attempting an interpretation, others are less convinced especially because of the ad hoc choices of the simplicity functional. In general, rotated EOF are more stable than the conventional vectors since they introduce another constraint that can be used to distinguish between eigenvectors. The well separated rotated EOF are therefore more resilient and then show less sensitivity to the errors that we have discussed in the previous chapters.

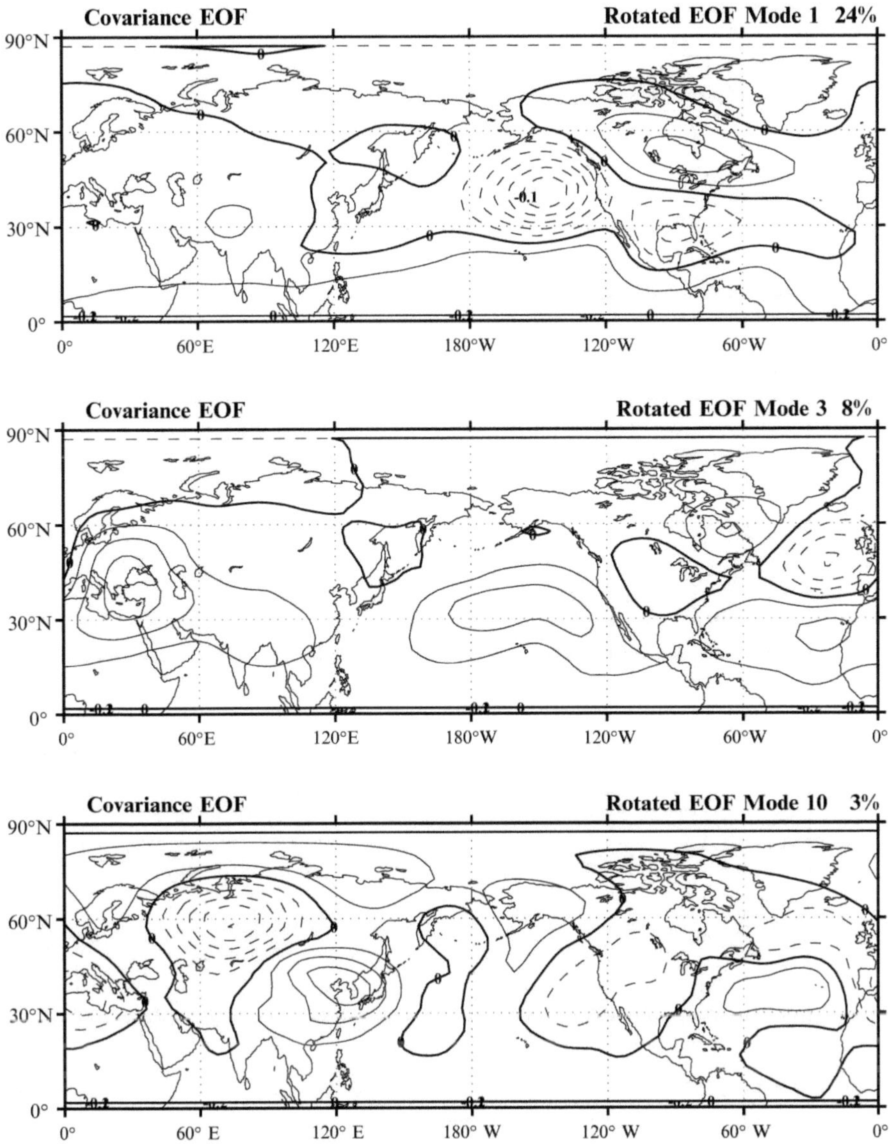

Fig. 5.2 Conventional EOF for the test data sets Z500

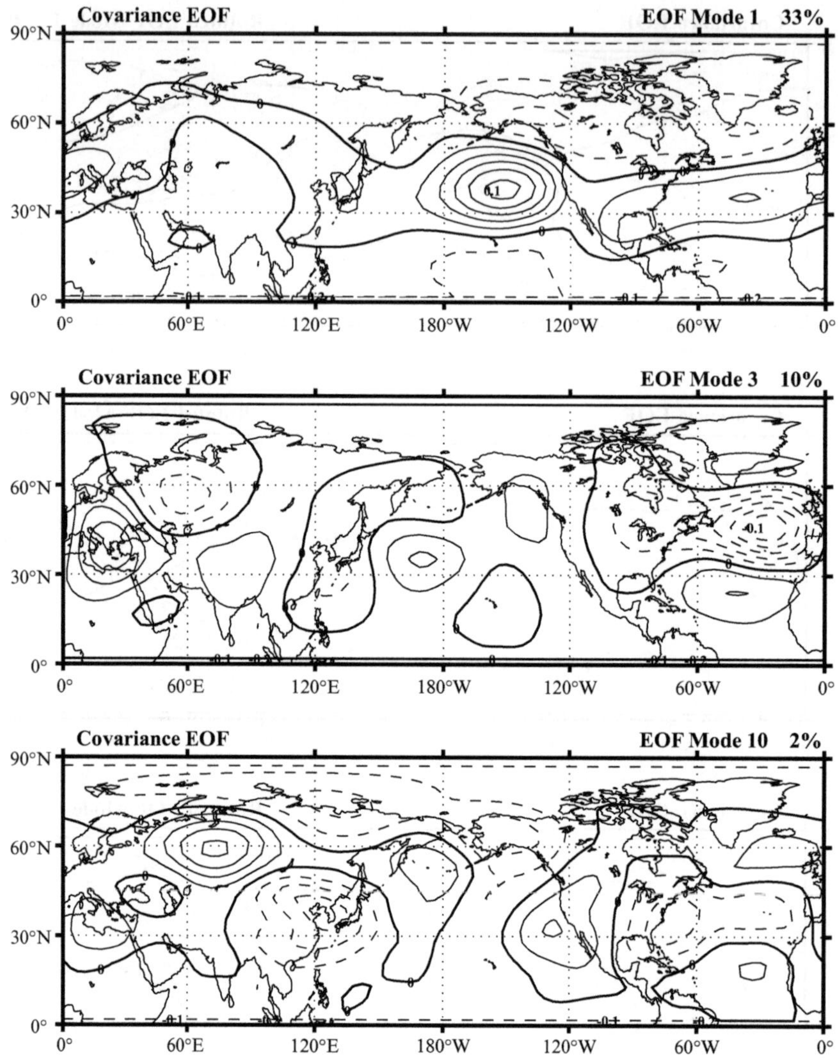

Fig. 5.3 Rotated EOF according to the normal VARIMAX method for the test data sets Z500. Also shown is the variance explained by the rotated mode

Exercises and Problems

1. Given the set of data

$$\mathbf{X} = \begin{pmatrix} 1 & 1 & 0 & 1 & 0 \\ 1 & 0 & -1 & 0 & -1 \\ -1 & 1 & 0 & 1 & 0 \\ 1 & 0 & 1 & -1 & 1 \\ -1 & 1 & 0 & 1 & 1 \\ 0 & 0 & -1 & 0 & -1 \\ -1 & 1 & 1 & 1 & 1 \\ -2 & 1 & 1 & 1 & 1 \end{pmatrix},$$

compute the first two EOF and the rotated EOF with VARIMAX (use the MATLAB functions `princomp` and `rotatefactors`).

The command `L = princomp(X);` *yields*

$$\boldsymbol{L} = \begin{pmatrix} -0.6756 & -0.4171 & 0.6047 & -0.0192 & 0.0600 \\ 0.2740 & 0.1259 & 0.4752 & 0.0060 & -0.8266 \\ 0.3931 & -0.5676 & 0.0620 & 0.7157 & 0.0847 \\ 0.3233 & 0.4582 & 0.6261 & 0.0680 & 0.5374 \\ 0.4577 & -0.5273 & 0.1125 & -0.6948 & 0.1310 \end{pmatrix},$$

and the subsequent command `[L1,T]=rotatefactors(L(:,1:2));` *gives*

$$\boldsymbol{L}_1 = \begin{pmatrix} -0.7908 & 0.0714 \\ 0.2948 & -0.0636 \\ -0.0259 & -0.6900 \\ 0.5335 & 0.1728 \\ 0.0500 & -0.6964 \end{pmatrix}.$$

After rotation, it is possible to better decompose the data variance among the two principal components: the first data column and also somehow the forth column, are well represented by the first component. On the other hand, the third and fifth data columns are well represented by the second (rotated) principal component.

Non-orthogonal Rotations

The main conceptual difficulty with rotations is the fact that again we are forcing a condition on the data that we do not know whether it is reasonable to enforce. On the other hand the freedom of changing coordinate system includes transformations of type (5.3) that are not orthogonal, therefore we can ask whether it is possible to use a modal decomposition that does not require orthogonality from the start. By removing the orthogonality constraint, we are left with a large selection of possible transformations.

The method aims at identifying a transformation of a preliminary standard EOF pattern to achieve simpler structure. The transformation matrix is obtained by solving an oblique Procrustes problem. This mathematical problem can be stated as follows: Given matrices **A** and **B** of size $n \times m$ with **A** having full column rank, find a matrix **T** satisfying

$$\mathbf{B} = \mathbf{AT} + \mathbf{E}$$

such that the Frobenius norm of the error matrix **E**,

$$\|\mathbf{E}\|_F^2 = \text{trace}((\mathbf{B} - \mathbf{AT})^*(\mathbf{B} - \mathbf{AT})), \tag{5.4}$$

is minimized. **B** is often called the *target* matrix. The matrix **T** can be found as the only critical point of the function to be minimized, that is, as the solution of

$$\frac{\partial}{\partial \mathbf{T}}(\|\mathbf{E}\|_F^2) = -2\mathbf{A}^*\mathbf{B} + 2\mathbf{A}^*\mathbf{AT} = 0.$$

Solving for **T** yields

$$\mathbf{T} = (\mathbf{A}^*\mathbf{A})^{-1}\mathbf{A}^*\mathbf{B}.$$

The interpretation of the problem is relatively simple. The successful solution of the Procrustes problem is the identification of a linear relation between two sets of data. In case **A** is not full column rank, $\mathbf{A}^*\mathbf{A}$ is singular and **T** cannot be determined as outlined above. However, a (non-unique) minimizing solution can always be obtained by recurring to the pseudoinverse of $\mathbf{A}^*\mathbf{A}$ (cf. end of Sect. 2.8).

The PROMAX method uses the Procrustes problem to obtain a simple structure solution. The basic idea is to create a "simple" target matrix and then use a Procrustes transformation to obtain an oblique set of modes that have a more insightful structure than the original modes. The observation that orthogonally rotated modes, such as those obtained by VARIMAX, are usually a good deal simple themselves suggests that the VARIMAX modes can be used as starting point. Therefore, each element of the target matrix **B** can be defined as

$$b_{ij} = \frac{|v_{ij}|^k}{v_{ij}}, \tag{5.5}$$

where v_{ij} are the VARIMAX pattern values in each spatial point, previously normalized. The Procrustes problem is then formulated with **B** as target matrix and with **V** the VARIMAX pattern matrix as the data matrix

$$\mathbf{B} = \mathbf{VT} + \mathbf{E},$$

with solution

$$\mathbf{T} = (\mathbf{V}^*\mathbf{V})^{-1}\mathbf{V}^*\mathbf{B}.$$

The oblique patterns are then given by

$$\mathbf{V}_{\text{promax}} = \mathbf{VTD},$$

where the matrix **D** scales the oblique modes to unit length, namely

$$\mathbf{D}^2 = \text{diag}(\mathbf{T}^*\mathbf{T})^{-1}.$$

The definition of the target matrix as a power of the original pattern (cf. the exponent k in (5.5)) is an attempt to emphasize the differences between maxima and minima, to obtain a simpler structure in which intermediate values are unfavored. The value of the parameter k is arbitrary, but there is a difference in the sensitivity of the modes according to the shape of the sought after real pattern. If we expect a strong pattern with large variations between its extreme values, then k should be set to a low number. In practice, $k = 2$ or $k = 4$ are often used.

The comparison between the standard and rotated modes is shown in Fig. 5.4 for the first mode in the test data set for Z500. The orthogonal VARIMAX rotation results in an intense pattern, better localized, as we have seen in the preceding pictures. The PROMAX solution (lower panels) obtains patterns even more localized on North America, but we can notice one of the problems with PROMAX, especially if a large value of k is selected (bottom panel is for $k = 12$). The construction of these modes tends to polarize the spatial variability, concentrating the variance in smaller regions. The modes have fewer peaks, but of larger amplitude. We can see that for $k = 12$ the centers are more intense, even in regions where the EOF or the VARIMAX showed little amplitude. This example emphasizes that simple structure in principle that does not necessarily imply more meaningful modes. Figure 5.1 shows that from the point of view of simple structure, i.e. the polarization and separation of the pattern values in space, we are getting better every time. The concept of simple structure is therefore a very useful concept, but it cannot be considered as the only guiding principle.

Oblique modes have not found a widespread usage in data analysis, perhaps because of the parametric freedom, but also because they cannot be used to separate the variance.

5.3 Complex EOF

We have seen how conventional and rotated EOF can be employed to identify patterns that optimize the explanation of the variance. EOF identify the dominant pattern, but the information on the time evolution is only implicitly included into the evolution of the coefficients. Data that contain oscillations in time or in space and time as a propagating signal, are very common in applications. In Sect. 4.5.1 we have seen an example in which the standard EOF have been applied to an ideal example of a propagating wave. The signature of the propagation is visible in the

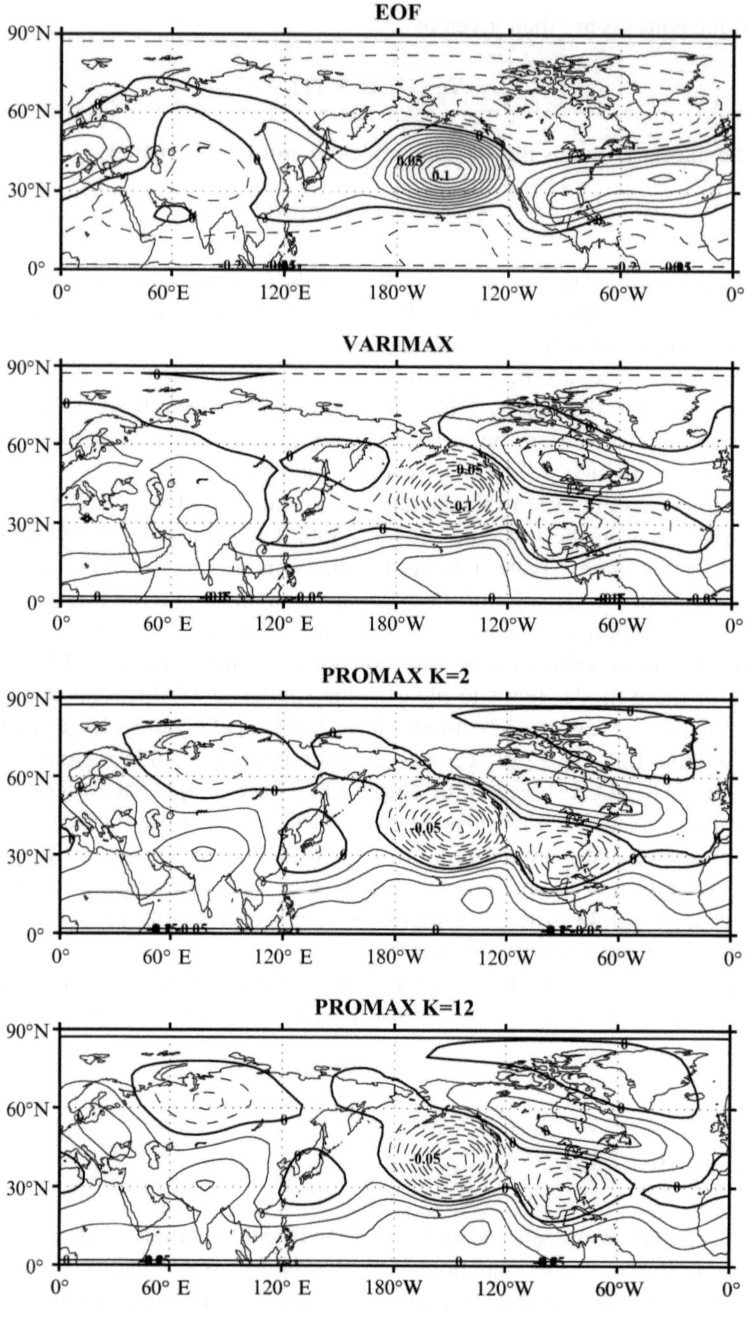

Fig. 5.4 Conventional, rotated and PROMAX (*oblique*) EOF for the test data sets Z500

EOF, but it requires some indirect interpretation. The presence of propagation is indicated by two modes whose patterns are in quadrature, namely the relative maxima and minima of one pattern correspond to the zero lines of the other and the two EOF explain a similar amount of variance (see Fig. 4.13).

The variations of the coefficients in time (top panel of Fig. 5.5) show a periodic behavior in time. There is a shift in time corresponding to a quarter of a wavelength between EOF1 and EOF2. A quarter wavelength shift in time is the phase lag typical of a harmonic wave of the form

$$V(\mathbf{x}, t) = \Re[U(\mathbf{x})e^{-i\omega t}] = \Re[U(\mathbf{x})(\cos(\omega t) + \imath \sin(\omega t))]. \quad (5.6)$$

Therefore, the variation in time of the EOF coefficient seems to identify a kind of variability that can be expressed as a harmonic wave with real part EOF1 and imaginary part EOF2. The EOF analysis has been able to find couples of modes that are strongly linked, in fact they may be part of the same physical system.

Waves are a pervasive physical phenomenon so it is not surprising that the EOF's feature of detecting propagating modes has raised considerable interest. On the other hand, it is also true that this capability is a sort of byproduct of the general property of EOF to maximize variance. Would it be possible to sharpen the EOF definition so as to go after propagating modes?

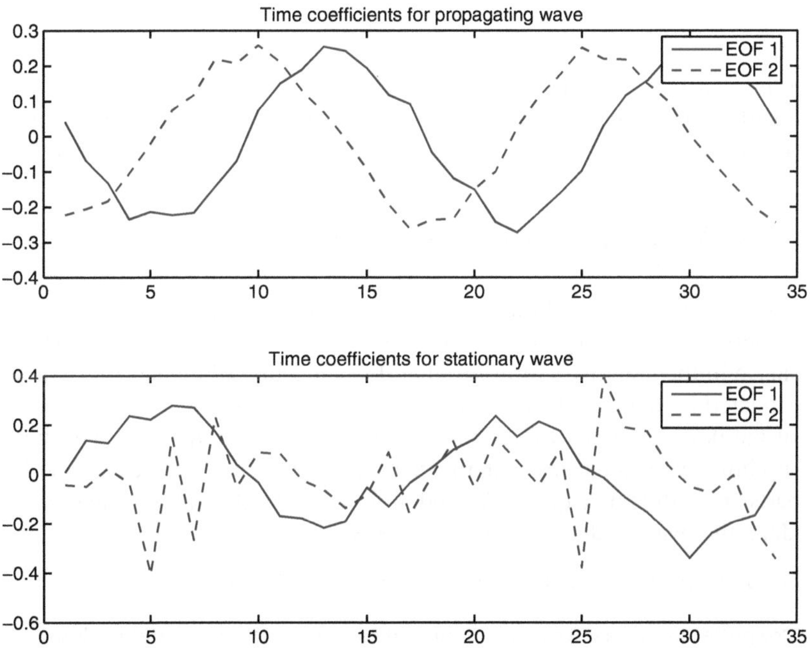

Fig. 5.5 EOF coefficients of the example in Sect. 4.5. *Top panel*: a propagating wave. *Bottom panel*: a stationary wave

We have seen that the quarter wavelength shift is a peculiar phase relation that indicates propagation. Can we find a way to enhance the modes that are in that particular phase relation? One possibility is to change the available data to stress the phase relation we are looking for; in our case we can expand the data by adding a new data set obtained by shifting all data by one quarter wavelength. This is a mathematical procedure that can be performed by *Hilbert transform*. The analytical definition of the transform is

$$\hat{f} = H[f(x,t)] = \frac{1}{\pi}\int_{-\infty}^{\infty} \frac{s(\tau)}{t-\tau} d\tau$$

where the integral is to be understood to be a Cauchy principal value to avoid the singularities at infinity and at $t = \tau$. In practice, the transform of discrete signal is performed using a discrete Fourier transform (Hahn 1996)

$$\hat{f} = H[f(x,t)] = \sum_{\omega} f_H(x,\omega) e^{-2\pi\iota\omega t}, \quad f_H(x,\omega) = \begin{cases} ig(x,\omega) & \text{for } \omega > 0 \\ 0 & \text{for } \omega = 0 \\ -ig(x,\omega) & \text{for } \omega < 0. \end{cases}$$

where $g(\omega)$ is the discrete Fourier transform of f. The Hilbert transform shifts the data series a quarter period to obtain a new, augmented, data series of complex data,

$$\mathbf{X}_C = \mathbf{X} + iH(\mathbf{X}),$$

where the real part contains the original data and the imaginary data the quarter period shifted data. Let us assume that $\mathbf{X}_C$ has been detrended, so that its mean is zero. The variance is thus given by the sum of the diagonal elements of the following matrix

$$\mathbf{X}_C\mathbf{X}_C^* = \mathbf{X}\mathbf{X}^* + H(\mathbf{X})^* H(\mathbf{X}) + i(\mathbf{X}(H(\mathbf{X}))^* - H(\mathbf{X})\mathbf{X}^*). \quad (5.7)$$

Therefore, the variance of the new data set $\mathbf{X}_C$ is twice the variance of the original data series, as the imaginary term does not contribute to the variance. However, the balance in the imaginary term is rather delicate and it often happens that in real cases affected by noise, the variance is only *approximately* twice the original variance of the data. Complex EOF defined through Hilbert transforms will therefore try to optimize variance using patterns that are complex and whose real and imaginary parts are shifted by a quarter period. Below is a Matlab implementation of this procedure, that was used to generate later plots.

```
function [u,lam,v,proj]=ceof(z,indf,nmode,nproj)
%
% Compute complex EOF of z and expand it for nmode modes
%
% Inputs:
%         z Data Matrix
%         indf        Index for the data
```

```
%        nmode       Number of EOF to return
%        nproj       Number of EOF to generate projections
% Outputs:
%        u           EOF arrays (nspace x nmode)
%        lam         variance explained    (ntime)
%        v           Unnormalized EOF coefficients
%        proj        Projection on the nmode EOF
%
resol = [96 48];
zh=hilbert(z);
[uu,ss,v]=svd(zh,0);

lam = diag(ss).^2/sum(diag(ss).^2);     % Explained variances

u=zeros([resol(1)*resol(2) nmode]);     % Keep Only first modes
u(indf,1:nmode)=uu(indf,1:nmode);
proj=zh'*uu(:,1:nproj);     % Compute projections

return
```

Figure 5.6 shows the first complex EOF (CEOF 1) for the case of the analytical wave of Sect. 4.5.1. The top panels show the real and imaginary parts of the first mode and they display the familiar shape in quadrature one with the other. The real and imaginary parts of the coefficient are also shifted one quarter wavelength. We can see that the CEOF has recovered the propagating wave hidden in the noise.

Being focused on extracting the signals that are shifted one quarter wavelength, the CEOF are very efficient at doing that, but at the same time the Complex EOF do not comparably perform if the oscillatory signal has a structure with a different phase relation. For instance, if the signal is stationary, namely it changes in time without a change of phase in space, like an oscillating beam, CEOF run into trouble. Propagation and stationarity are identified clearly in our ideal experiment by simple EOF (Fig. 5.5) because the stationary signal (bottom panel) shows no clear phase relation between the time series of the coefficient. Application of the CEOF to a stationary signal (Fig. 5.7) produces a spatial pattern that bears indication of the signal stationary nature. Only the real or imaginary component is now needed to give the spatial structure of a stationary signal, in this case the real part, whereas the other component is usually noise, without a clear pattern. It would appear that CEOF have successfully identified the signal, however if one looks at the time coefficient (bottom panel) it is possible to see that both time coefficients oscillate, pretty much in the same way as in the preceding propagating case. CEOF can only distinguish between spatial propagation and lack of it, implying the absence of spatial phase relations; in general, however, the inspection of the time coefficient alone is not sufficient to distinguish between them. As an example, in Fig. 5.6 it is possible to see that the variation of the spatial phase (the arrows in the panel) is organized and smooth, corresponding to the organized propagation. In contrast, in Fig. 5.7 the phase variation is disorganized and dominated by noise. This investigation can be somewhat difficult to perform with real data, where spatial phase relations are

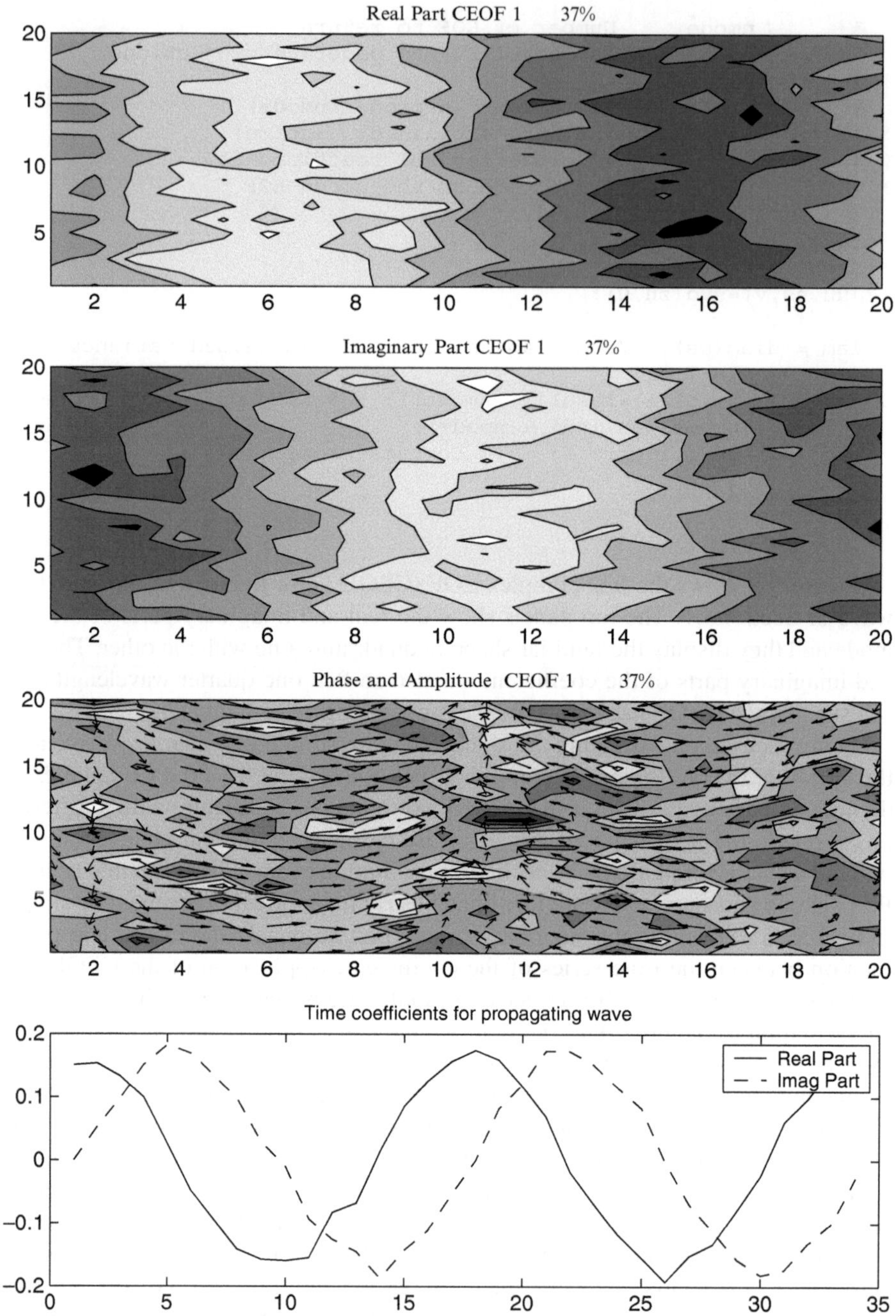

Fig. 5.6 First complex EOF of the analytical example. *Top panels*: spatial patterns of the real and imaginary parts, then the amplitude and phases of the mode. In the title, the explained variance is recorded. *Bottom panel*: time evolution of the coefficient

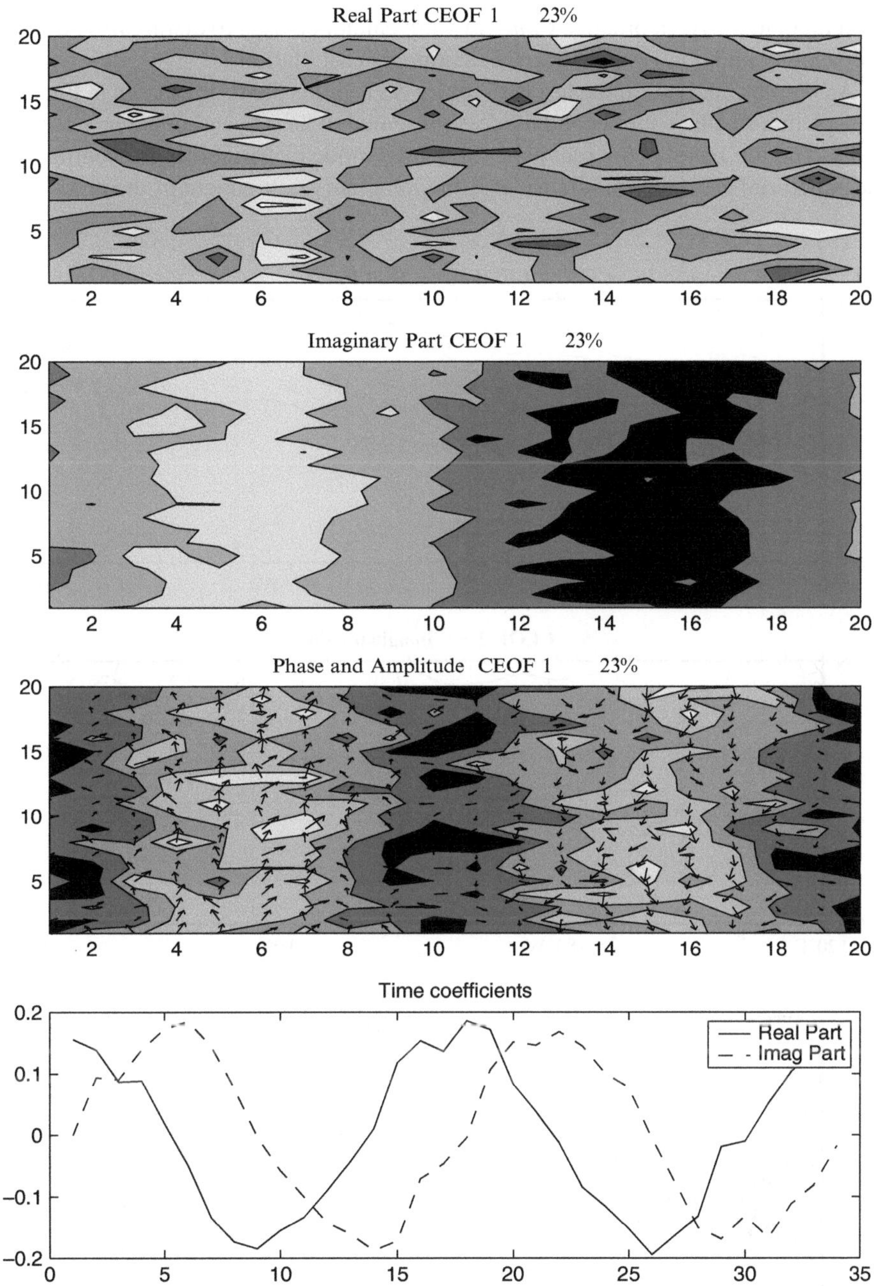

Fig. 5.7 As in Fig. 5.6 but for the case of a stationary wave

difficult to identify. In practice *Complex EOF* cannot be used to distinguish between propagating and non-propagating (i.e. stationary) oscillations.

A complex analysis of the test data set for SST yields the result shown in Fig. 5.8. This picture displays the second mode represented in its real and imaginary components. The top panel is the real part, showing a mode of variability concentrated in the equatorial area, the middle panel is the imaginary component of the mode. The

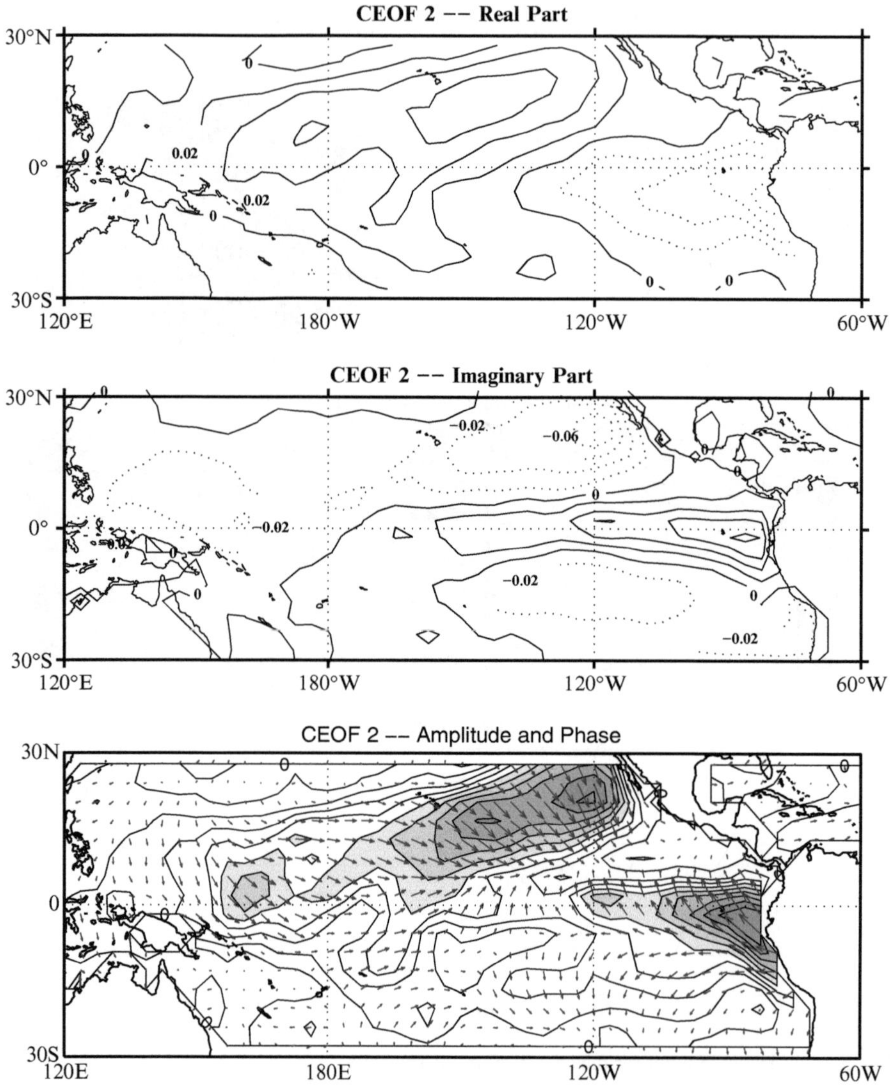

Fig. 5.8 Second complex EOF of the marine temperatures in the Pacific. *Top panel*: real component, *middle panel*: imaginary component, *lower panel*: amplitude and phase

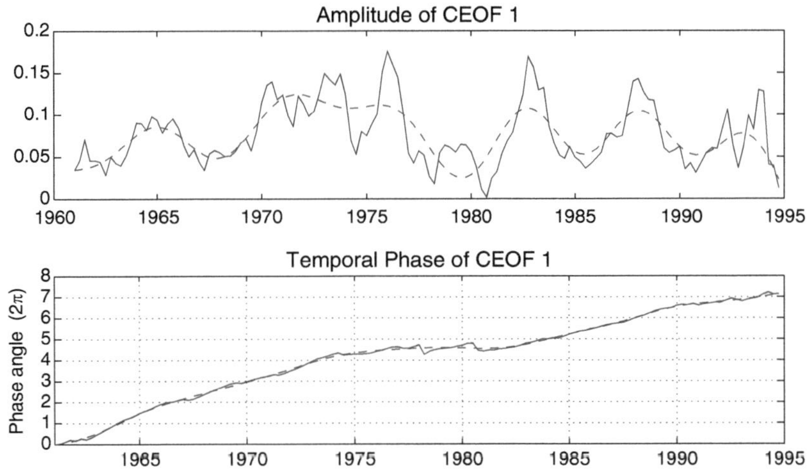

Fig. 5.9 Time series of the coefficient of the second Complex EOF shown in the previous picture. The *top panel* shows the amplitude of the complex coefficient, whereas the *bottom panel* shows the evolution of the unwrapped phase angle. The phase velocity is obtained as the derivative of the phase, showing an acceleration after 1980

bottom panel is the representation in amplitude and phase. The amplitude is concentrated in the east equatorial Pacific, the rotation of the phase indicates a phase velocity towards the west. Here the convention used is that the phase arrows point to the east if the real part is positive and the imaginary part is zero.

The time evolution of the mode coefficient is displayed in Fig. 5.9, indicating the periods of time in which such a mode is more or less energetic. The "unwrapped" phase, that is the phase of the time coefficient reduced to a single-value function by adding a factor 2π every time it crosses the zero line, also shows different phase speed from a period to the next.

Figure 5.10 shows the modal actual evolution, cycling through the real and imaginary parts with alternate signs. The reported field is only based on the reconstruction of the second mode, starting from 1980 onward. The picture shows that the CEOF indicates an oscillatory behavior that can also be aperiodic in time. There are periods in which oscillations are clearly visible, and periods where oscillations are quiescent and there is very little appearance of the mode. This is a good example of the capability of the CEOF to capture irregular oscillations.

5.4 Extended EOF

Complex EOF are based on the analysis of variance by taking into account the data time behavior. This is done by creating a new data set that includes the original data series and a new series that is shifted by a quarter wavelength. The Hilbert transform

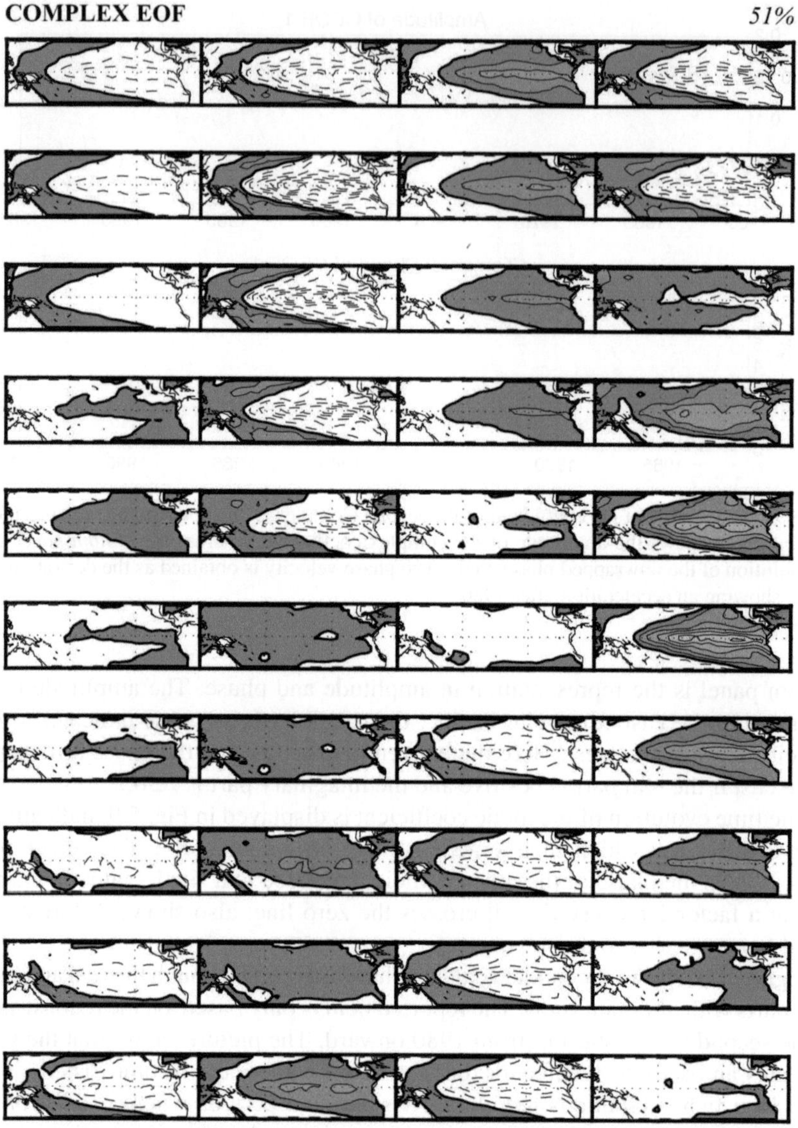

Fig. 5.10 Time evolution of the second for SST from Winter 1980 (*top left panel*) for each consecutive season. Time is increasing downwards and from left to right. The 1982–1983 El Niño event is visible in the first and second column on the left

makes the procedure very rigorous. However, it is sometimes desirable to use a less rigorous approach and to gain some flexibility in the process. *Complex EOF* change the state vectors in a way that the basic data are not the data at a given time, but the combination data at a single time plus data shifted one quarter wavelength in time. A possible alternative is to introduce a derivative EOF analysis that crudely realizes this fact. This new method, often called the Extended EOF (EEOF), simply consists in extending the data set with repetitions of the time series suitably lagged. For the test cases we are using here it will mean to extend the data by adding several copies of the time series with proper time shifts, i.e.

$$\mathbf{X}_E = \begin{bmatrix} \mathbf{x}_1 & \mathbf{x}_2 & \cdots & \mathbf{x}_{m-2} \\ \mathbf{x}_2 & \mathbf{x}_3 & \cdots & \mathbf{x}_{m-1} \\ \mathbf{x}_3 & \mathbf{x}_4 & \cdots & \mathbf{x}_m \end{bmatrix}.$$

The basic observation vector at time n is given by

$$y_E(n) = \begin{bmatrix} \mathbf{x}_n \\ \mathbf{x}_{n+1} \\ \mathbf{x}_{n+2} \end{bmatrix}.$$

It is formed by $k+1$ fields, each showing the dominant mode of variations over the k lags. A single mode is then formed by several components each representing the spatial pattern for that phase of the lags. The trick is to include the lags that are important for reproducing possible oscillatory patterns. It is advisable to investigate the autocorrelation function to gather some indications of the number of lags that need to be included. The method is very flexible, the lags do not need to be consecutive. Instead of using three consecutive months like in the previous example, we could have chosen some three months in three months. In principle they do not even need to be equally distributed; arbitrary lags could be defined, but results would be extremely difficult to interpret. In practice it is advisable to use regularly spaced lags. The variance of the augmented series is a multiple of the variance of the original series and it is approximately $k+1$ times the original variance, so the amount of variance explained must be assessed against this augmented variance.

A simple Matlab implementation of the Extended EOFs approach follows.

```
function [u,lam,v,proj]=eeof(z,indf,nmode,nproj)
%
%Compute Extended EOFs of matrix z and expand it for nmode modes
% Use  3 lags
% Inputs:
%          z Data Matrix
%          indf    Index for the data (from the reading routine)
%          nmode   Number of EOF to return
%          nproj   Number of EOF to generate projections
% Outputs:
%          u       EOF arrays (nspace x nmode)
%          lam     variance explained    (ntime)
%          v       Unnormalized EOF coefficients
%          proj    Projection on the nmode EOF
%
resol = [96 48];
[np,nt]=size(z);
```

```
lags=3;      nmode=2;
zh=ones((lags+1)*np,nt-lags);
zh(1:np,:) = z(:,1:nt-lags);
zh(np+1:2*np,:) = z(:,2:nt-lags+1);
zh(2*np+1:3*np,:) = z(:,3:nt-lags+2);
zh(3*np+1:4*np,:) = z(:,4:nt-lags+3);
[uu,ss,v]=svd(zh,0);

lam = diag(ss).^2/sum(diag(ss).^2);   % Explained variances

u=zeros([resol(1)*resol(2) 4]);        % Only first mode
uc=zeros(np,4);
for i=1:4
    uc(:,i) = uu((i-1)*np+1:i*np,nmode);
end
u(indf,1:4)=uc(:,1:4);

proj=zh*uu(:,1:nproj);      % Compute projections

return
```

The example reported in Fig. 5.11 shows the result of applying an EEOF analysis to the tropical SST. The lags have been defined to the seasonal means of the SST and three seasonal lags have been used. It is possible to see how the main pattern of variations are captured.

Exercises and Problems

1. Show that the diagonal terms of the imaginary term in (5.7) do not contribute to the variance of the field.
2. Show that the total variance of the EEOF time series is approximately $k + 1$ times the original one, and that the approximation gets better as the number of time observations increases.
3. Construct the time evolution for the first mode for the EEOF technique.

5.5 Many Field Problems: Combined EOF

The extension of the EOF analysis to the time domain shows that the concept is more general than we may have thought. The logical path that we have followed to go into the time domain has exploited the freedom to change the rules of compositions of the data fields. We have generated other ways to analyze variance by arranging/transforming the data differently. The extension we have made in the previous section was mainly in the time variable, but we can use the freedom to change the definition of the data vectors to explore the variation of combined fields. We can, for instance decide to define a new data set by putting together the height and SST

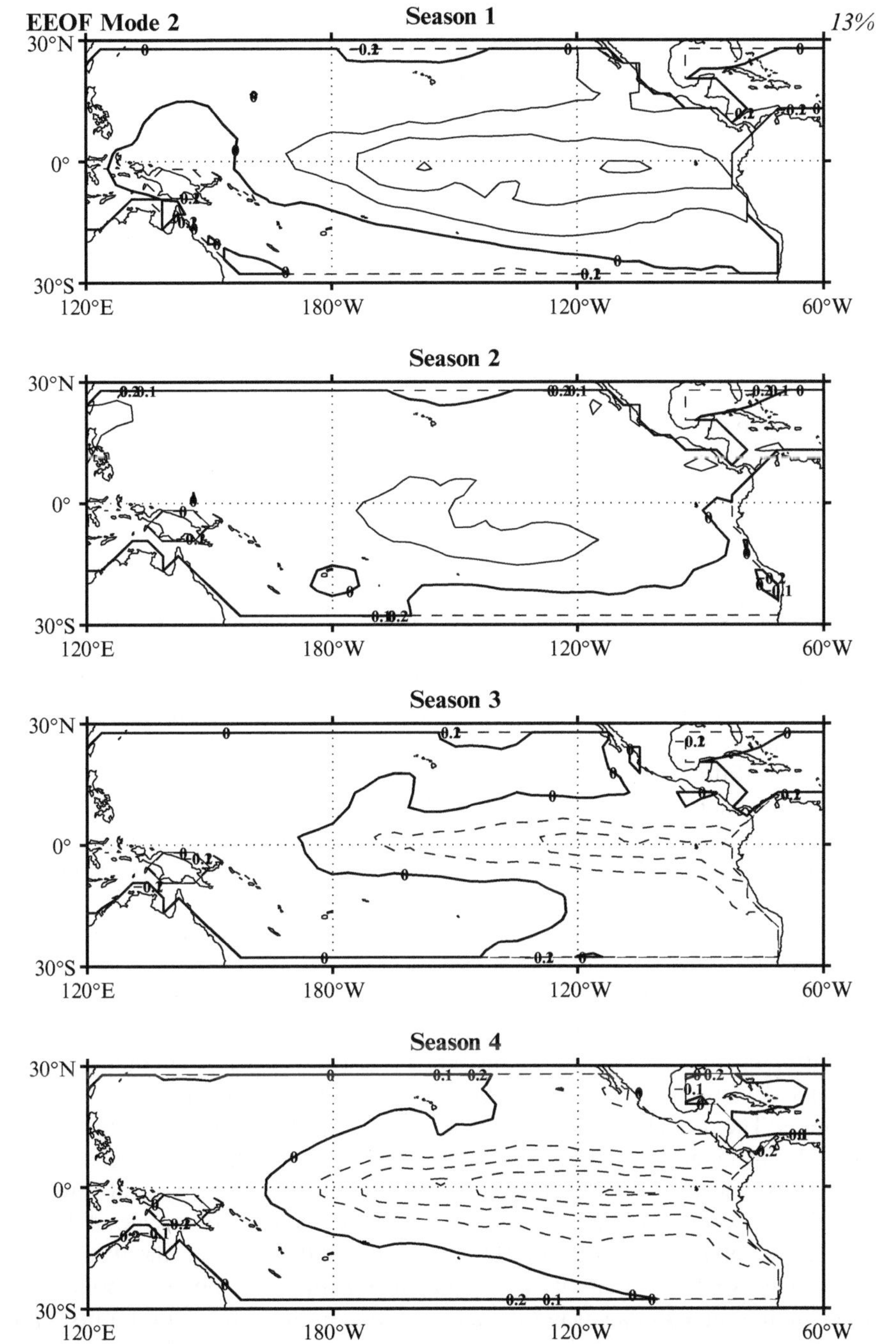

Fig. 5.11 First EEOF mode for the SST data set. The analysis has been performed by season, using three lags of one season each. The picture depicts the evolution of the mode through four consecutive seasons. The amount of variance explained by the mode is referred to the total variance of the augmented series

data. The data matrix can then be written as

$$\mathbf{Y}_n = \begin{bmatrix} \mathbf{z}_1 \cdots \mathbf{z}_n \\ \mathbf{s}_1 \cdots \mathbf{s}_n \end{bmatrix},$$

where the data are arranged in such a way to keep the time correspondence between the different fields, so that fields at the same time are put in the same column. We can also use the data matrix for the fields $\mathbf{Z} = [\mathbf{z}_1, \mathbf{z}_2, \ldots, \mathbf{z}_m]$ and $\mathbf{S} = [\mathbf{s}_1, \mathbf{s}_2, \ldots, \mathbf{s}_m]$ so that the new combined data matrix Y becomes

$$\mathbf{Y} = \begin{bmatrix} \mathbf{Z} \\ \mathbf{S} \end{bmatrix}.$$

Assuming zero mean, we can compute the covariance matrix for the combined field as

$$\mathbf{YY}^* = \begin{bmatrix} \mathbf{Z} \\ \mathbf{S} \end{bmatrix} [\mathbf{Z}^*, \mathbf{S}^*] = \begin{bmatrix} \mathbf{ZZ}^* & \mathbf{ZS}^* \\ \mathbf{SZ}^* & \mathbf{SS}^* \end{bmatrix}, \tag{5.8}$$

showing that the total variance of the combined field is the sum of the variance of the composing fields.

The two data sets can have different geographic extensions, though they must have the same number of time levels. There is also no limitation in the number of fields that are patched together in this way. We can put in the same data space three or four different fields, in principle there is no limit. This a very useful and rather unique feature of the combined EOF. There are several situations when this may be convenient. For instance, when treating tropical air-sea phenomena it is often useful to look for combined modes of variations of wind stress, SST, Outgoing Lonwave Radiation (OLR), precipitation, clouds, etc. The combined EOF is the only method that allows a simultaneous considerations of the possible modes of variation of different variables.

The combination of fields in this way requires some care to handle different units and quantities. Different data have widely different numerical values corresponding to the different units that are used to measure them. These differences could generate systematic deviations in the resulting patterns that do not correspond to real variability patterns. The problem can be overcome by transforming the data to values of the same order of magnitude by using suitable scales, making the data adimensional. The simplest way is to divide the data by constants that represent typical value for that variable. For instance, in our case we could use a temperature scale of 300 K, and a geopotential height scale of 5000 m, that would change all the data values to order one. Another possibility is to normalize them by the point-by-point standard deviation, in a similar way to what was done in Sect. 4.4.1. In the first case the scaling is simply equivalent to a multiplication by a constant and the covariance structure is not modified, so we get the Combined Covariance EOS, in the latter case the covariance structure is modified and we get Combined Correlation EOF.

Each mode is now a combination of the fields that have been used to create the combined data set. The mode describes the principal mode of variations of the combined data and it is not different from the EOF that we have described in the previous chapter. However, the various fields can be identified in the mode by reconstructing the different components with the corresponding order in the data field. In this sense, the combined EOF is a straight generalization of the EOF that can be considered as a one-parameter Combined EOF. A typical implementation is as follows

```
function [u,lam,v,proj]=combeof(zz,inds,indz,nmode,nproj)
%
% Compute combined EOF of matrix zz. The matrix zz contains
% the ordered fields to be combined, in this case Z and S.
% Inputs:
%           zz    Combined Data Matrix
%           inds       Index for the S data (ocean)
%           indz       Index for the Z data (atmosphere)
%           nmode      Number of EOF to return
%           nproj      Number of EOF to generate projections
% Outputs:
%           u          EOF arrays (nspace x nmode)
%           lam        variance explained    (ntime)
%           v          Unnormalized EOF coefficients
%           proj       Projection on the nmode EOF

resol = [96 48];   ss=resol(1)*resol(2);
[uu,ss,vv]=svd(zz,0);
lam = diag(ss).^2/sum(diag(ss).^2);    % Explained variances

ls=length(inds);        lz=length(indz);
u=zeros([ss nmode]);      v=zeros([ss nmode]);
for i=1:nmode
    u(indz,i)=uu(1:lz,i);
    v(inds,i)=uu(lz+1:lz+ls,i);
end

proj=zz*uu(:,1:nproj); % Compute projections

return
```

We have used the standard deviation normalization to produce Combined Correlation EOF of the height and SST fields showed in Figs. 5.12 and 5.13. The mode resembles very much the EOF obtained by performing the analysis of the SST or the height field alone. The pattern can be superposed almost exactly. It is possible to understand this effect by inspecting the structure of the combined data covariance matrix in (5.8). The structure is essentially given by a block matrix structure where the blocks are the covariance matrix of the component fields along the diagonal and the cross-covariance matrices of the fields in the off diagonal positions. Therefore, the diagonal terms express the internal variability of the fields, whereas the off-diagonal terms express the variance of one field that is related to the other field.

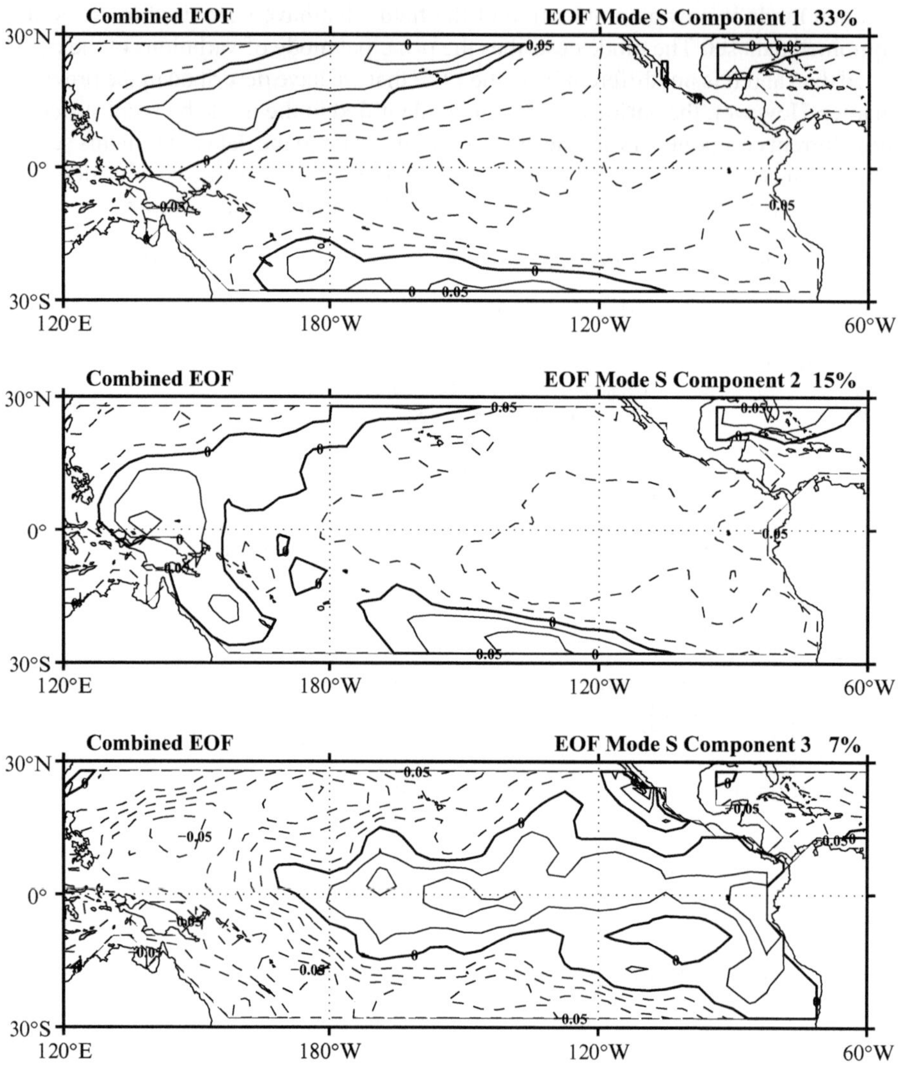

Fig. 5.12 The first three combined EOF modes for the Height-SST data set. Here is shown the SST component in descending order of explained total combined variance

The combined EOF will obtain the same EOF as the individual fields if the off-diagonal terms are small compared to the diagonal ones. This happens if the data fields are independent of each other and therefore the cross-covariance components are small; in this case, the structure of the combined covariance matrix is essentially dominated by the individual covariance of the fields. The combined EOF will be dominated by the autocovariance of each field if the internal variability of the fields less larger than the cross-covariance.

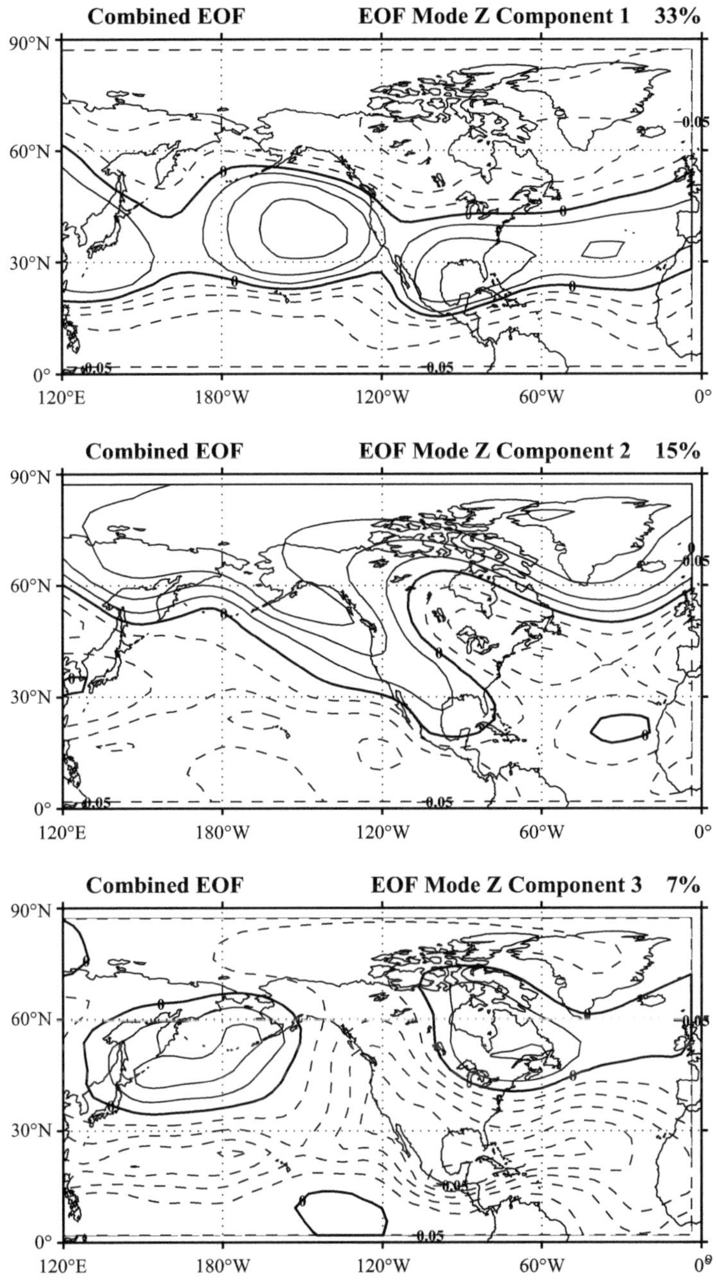

Fig. 5.13 The first three combined EOF modes for the Height-SST data set. Here is shown the Z component in descending order of explained total combined variance

This observation leads to the main weakness of the combined EOF: by mixing the autocovariance of each field and the cross-covariance of one field with the other, combined EOF cannot separate the patterns for the different kind of variability and one cannot tell the respective amount due to the autocovariance or to the cross-covariance. The Combined EOF mode will bear the imprint of both sectors of variability of a particular variable. It is a pity, because the cross-covariance could be extremely useful when one has to study coupled problems, like the air-sea interaction in the tropics. The Combined EOF cannot give a suitable help on this issue, but we will see in the following chapter that we can work out specific methods to address this exciting issue.

Chapter 6
Cross-Covariance and the Singular Value Decomposition

6.1 The Cross-Covariance

At the end of the previous chapter we have introduced the concept of the simultaneous analysis of different fields. We have introduced the Combined EOF that, after a suitable scaling, allow us to produce patterns of variability that reflect the covariance properties of different data types. This is an interesting development because it leads to the consideration of the cross-covariance along the same lines we have used for the covariance of a single field. The program we have followed in Chaps. 4 and 5 has been inspired by the attempt to analyze the variance of a single field, finding the best way to represent the data, maximizing the variance with the smallest number of patterns. The modes we have found have been identified as "preferred" modes of variations and we have shown that they are linked to the number of degrees of freedom in the data space.

The observations on the cross-covariance poses now the question whether it is possible to proceed with an analogous program for the cross-covariance. Can we analyze the cross-covariance in a similar way as what we have done for the covariance? Can we identify patterns that are "preferred" in the sense that few of them can explain most of the cross-covariance between two fields? The Combined EOF seems a good conceptual starting point, but we will have to abandon the extra flexibility to allow for any number of fields restricting the analysis to two fields.

Starting from the definition of the data space that we have used for the Combined EOF, the data matrix can be written as

$$\mathbf{X} = \begin{pmatrix} z(1) & z(2) & \ldots & z(n) \\ s(1) & s(2) & \ldots & s(n) \end{pmatrix}.$$

As in the preceding chapter, data are arranged so that fields at the same time are in the same column. Introducing in a similar manner the data matrices for the fields

$$\mathbf{Z} = (\mathbf{z}(1), \mathbf{z}(2), \ldots, \mathbf{z}(n)), \qquad \text{and} \qquad \mathbf{S} = (\mathbf{s}(1), \mathbf{s}(2), \ldots, \mathbf{s}(n)),$$

A. Navarra and V. Simoncini, *A Guide to Empirical Orthogonal Functions for Climate Data Analysis*, DOI 10.1007/978-90-481-3702-2_6,

where the vectors can have different length ($\mathbf{z} \in \mathbb{R}^m$ and $\mathbf{s} \in \mathbb{R}^p$), the new combined data matrix $\mathbf{Y}$ is

$$\mathbf{Y} = \begin{pmatrix} \mathbf{Z} \\ \mathbf{S} \end{pmatrix},$$

and the covariance matrix for the combined field can be written as

$$\mathbf{YY}^* = \begin{pmatrix} \mathbf{Z} \\ \mathbf{S} \end{pmatrix} \begin{pmatrix} \mathbf{Z}^* & \mathbf{S}^* \end{pmatrix} = \begin{pmatrix} \mathbf{ZZ}^* & \mathbf{ZS}^* \\ \mathbf{SZ}^* & \mathbf{SS}^* \end{pmatrix}. \tag{6.1}$$

The structure of the combined covariance already contains a lot of information. The relative size of the four blocks in (6.1) tells us something about the possible linear relation between the fields. It is possible, for instance, that the off-diagonal blocks are smaller than the diagonal blocks. An extreme situation occurs when the fields are independent of each other. In this case, in the language of the vector spaces, the data space splits in two separated subspaces, each corresponding to one of the fields, and the covariance matrix is

$$\mathbf{YY}^* = \begin{pmatrix} \mathbf{ZZ}^* & \mathbf{0} \\ \mathbf{0} & \mathbf{SS}^* \end{pmatrix}. \tag{6.2}$$

The EOF of the combined data sets are the same as the EOF of the individual fields. In the general case of non-negligible off-diagonal blocks, the EOF of the combined fields differ from the EOF computed for the single fields.

Therefore, the computed combined EOF represent both the cross-covariance of the fields involved, but also the "autocovariance" of the fields themselves and in this sense they do not represent the pure cross-covariance relations. We could think the data matrix of a field as composed of two parts

$$\mathbf{Z} = \mathbf{Z}_a + \mathbf{Z}_c.$$

The "autocovariant" part $\mathbf{Z}_a$ is annihilated by the other field so that

$$\mathbf{Z}_a\mathbf{S}^T = \mathbf{0},$$

and the signals that appear in $\mathbf{Z}$ can be divided into a signal that does not correlate with $\mathbf{S}$ and a portion that does. If this splitting is possible, then we are interested in ways to identify the space spanned by $\mathbf{Z}_c$, corresponding to the main modes of covariation.

A possible solution to this problem can be obtained by considering the cross-covariance matrix

$$\mathbf{Z}_a\mathbf{S}^T.$$

It is not possible to use the techniques we have described previously to identify the main modes of variations. In general such matrices are not square, so it is impossible to apply the eigenmode/eigenvector analysis and diagonalize it. Even in case of a square matrix, this would be unsymmetric in general, therefore its eigenvectors would not be pairwise orthogonal. Generic matrices of this kind are difficult to treat and they appear to have no particular properties that can be exploited to our aims. However, in this situation we can probably fully appreciate the power of the results presented in Chap. 2 regarding the Singular Value Decomposition. Any matrix can in fact be decomposed with an SVD. In our case we can apply the SVD to the cross-correlation matrix,

$$\mathbf{ZS}^* = \mathbf{U\Sigma V}^*. \tag{6.3}$$

The orthogonal matrices composed of the column vectors $\mathbf{u}$'s and $\mathbf{v}$'s, $\mathbf{U} = (\mathbf{u}_1, \mathbf{u}_2, \ldots, \mathbf{u}_n)$, $\mathbf{V} = (\mathbf{v}_1, \mathbf{v}_2, \ldots, \mathbf{v}_n)$ yield bases in the $\mathbf{Z}$ and $\mathbf{S}$ data fields, respectively, and the modes are paired, sharing the same singular values. The fields can be reconstructed using the columns of $\mathbf{U}, \mathbf{V}$ with

$$z_k = \sum_{i=1}^{n} u_i \langle u_i, z_k \rangle \qquad \text{and} \qquad s_k = \sum_{i=1}^{n} v_i \langle v_i, s_k \rangle . \tag{6.4}$$

We can give an interpretation of the singular values and modes that is similar to the EOF: the decomposition corresponds to saying that the total cross-covariance is given by the singular value sum. The ratio of each singular value can be interpreted as the fraction of cross-covariance that can be attributed to that particular mode, as in the following relation:

$$\mu_i = \frac{\sigma_i}{\sum_{i=1}^{n} \sigma_i}. \tag{6.5}$$

6.2 Cross-Covariance Analysis Using the SVD

The cross-covariance seems a natural extension of the concept of the covariance (the autocovariance) that we have discussed in the previous chapters. It is then reasonable to wander whether it is possible to extend the concept of EOF to the more general case. We have to resort to the general idea that the EOF are the directions, in the data space, that explain most of the cross-covariance. We may ask to look for a similar pattern that performs the same role. It is easy to see that if we choose the basis of the left and right singular vectors defined in the preceding section we can get indeed what we are looking for. With (6.4) in mind, we expand the data over the right and left singular vectors, obtaining two matrices of expansion coefficients, $\mathbf{A}$ and $\mathbf{B}$, that represent the data in the new bases formed by the singular vectors,

$$\mathbf{Z} = \mathbf{UA}, \qquad \mathbf{S} = \mathbf{VB}.$$

The cross-correlation matrix has a very simple form in this basis, namely

$$\mathbf{Z}\mathbf{S}^T = \mathbf{U}\mathbf{A}(\mathbf{V}\mathbf{B})^T = \mathbf{U}\mathbf{A}\mathbf{B}^T\mathbf{V}^T.$$

Comparing this decomposition with that in (6.3), it follows that it must be

$$\boldsymbol{\Sigma} = \mathbf{A}\mathbf{B}^T. \tag{6.6}$$

This means that the basis of the singular vectors of the cross-covariance matrix is such that the matrix itself is diagonal. It can also be shown that the diagonal values on the left-hand side of (6.6) are the values of the time series covariance of the singular vectors coefficient and that such covariances are maximized. Therefore, the singular vectors represent patterns with maximum covariances between the time series. They diagonalize the cross-covariance matrix in the same way the EOF diagonalize the covariance matrix, yielding a special basis of unconnected patterns. The following pictures show the result of this analysis when it is applied to the already introduced height and SST fields. Figure 6.1 shows the patterns for the $\mathbf{Z}$ component of the SVD analysis, whereas Fig. 6.2 shows the SST component.

The interpretation of the explained variance needs some discussion. The "diagonalization" of the cross-covariance matrix indicates that it is possible to interpret the diagonal singular values as the contribution to the total cross-covariance for that particular mode, in case we define the total cross-covariance as the trace of the matrix $\boldsymbol{\Sigma}$, namely the sum of the singular values. The modes can be ranked according to the amount of explained cross-covariance (CC) as it is represented in Figs 6.1 and 6.2. The amount is the same for the separate component of the mode. However, when we consider the mode components as a basis for the height or the SST, there is no guarantee that any relation exists between the relative importance of the two modes and it is possible that we get different amount of variance explained by the two factors. The amount of total variance explained by the two components when they are considered separately (TC) is also shown in the pictures. In some cases they are similar and in others different, there is no relation that forces a particular amount of explained variance.

The patterns themselves show some resemblance to the pattern of the EOF of the height field (Fig. 4.8) and to the Combined EOF (Fig. 5.13). This is not too surprising, since we are dissecting the same variability, each time trying to stress different aspects of it. The difference is larger when we go to higher modes, as it should be expected.

As in the case of the EOF, the analysis yields patterns that are idealizations, in the sense that they do not represent any physically realized pattern, but patterns that correspond to an optimization criterion for the cross-covariance. The SVD has identified patterns of covariation between the two different fields, as it can be seen from the inspection of the time series of the coefficients (Fig. 6.3).

It is possible to have a measure of the method's ability to capture the covariations, by applying it to a data set of randomly chosen data. Figure 6.4 shows the result when the cross-covariance SVD method is applied to a random data set of the same dimension in time and space of the previous pictures. The random nature of the cross-covariance is very well expressed by the absence of any structure, in the sense

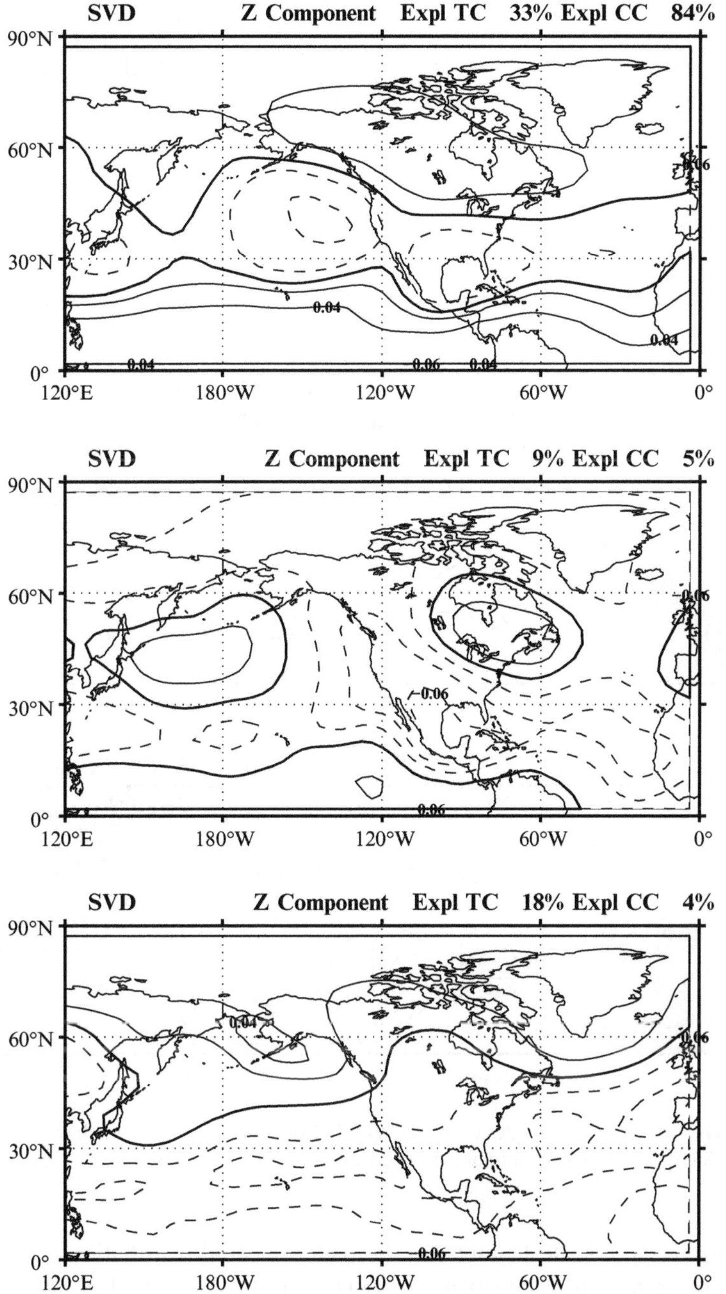

Fig. 6.1 The first three SVD modes for the Height-SST data set. Here is shown the **Z** component in descending order of explained total combined variance

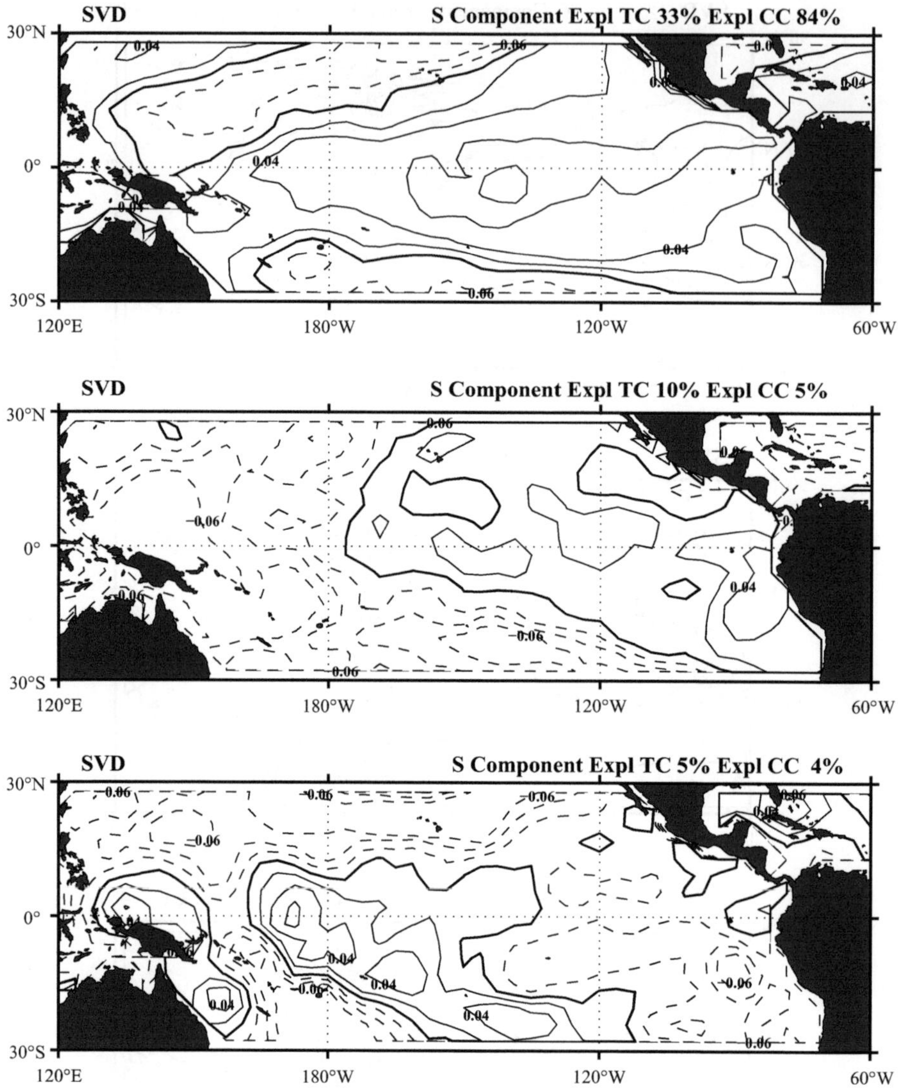

Fig. 6.2 The first three SVD modes for the Height-SST data set. Here is shown the SST component in descending order of explained total combined variance

of organized, large scale features, in the pattern spatial distribution. The patterns are indeed casual, indicating that the SVD is trying to do its best at optimizing the little cross-covariance occurring in the random data, but with limited success, since the amount of cross-covariance explained (TC) is very small, as the amount of total variance explained (CC) is also small. The cross-covariance is in fact almost

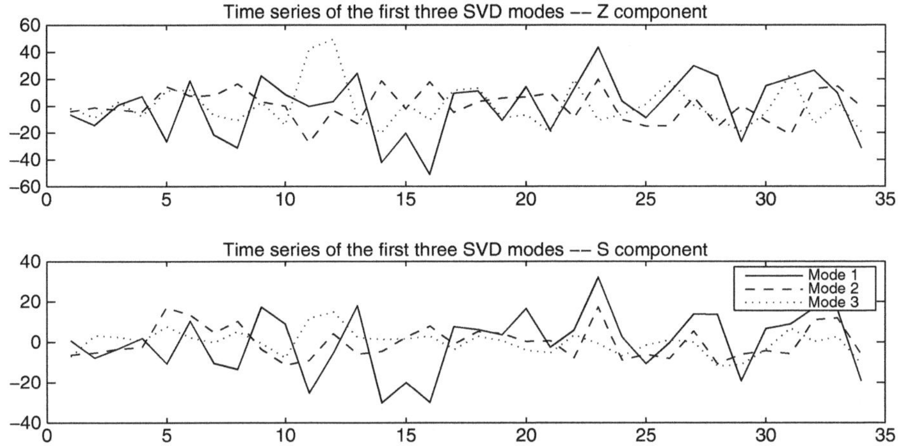

Fig. 6.3 The time series coefficients for the first three SVD modes for the Height-SST data set

uniformly distributed among the modes, a symptom of the data randomness and of the absence of any relation between the two fields.

Another interesting insight can be obtained by repeating the SVD analysis between the height and SST fields, but this time we rearrange the time series of the SST in a random way, so that we possibly destroy the relation in time that is the target of the SVD analysis. The permutation has been selected in such a way to reduce by 50% the amount of total cross-covariances between the two fields. It is totally arbitrary, so that months in **Z** in different years can be paired to different months of the SST. The results are shown in Figs. 6.5 and 6.6. The patterns do not show any particular deficiency and in fact they could be plausible patterns. They exhibit the kind of structure that can be considered typical of the variability of these fields, as we have seen in the various analyses so far. The SVD aims at emphasizing patterns that satisfy the optimization requirement even in presence of weak (by construction) cross-relation. Nevertheless the patterns themselves are plausible because they bring the signature of the internal variability (namely the individual covariance) of the fields so as not to get crazy patterns as in the previous example of the completely random fields.

This example shows one of the most insidious pitfalls of this kind of analysis. The mathematical requirements, the optimization criterion, the orthogonality constraint etc., can force structure even when it is unlikely to exist. In a real case, with no a priori knowledge of what structure to expect, there is a problem on how to assess the reliability of the structures found by the SVD and of the implicit covariant relations found. Actually, maybe some warning could be found in the small portion of total variance explained by the modes found by the SVD (see the TC values in the figures) that are in any case cautioning towards attributing an excessive importance to the structures.

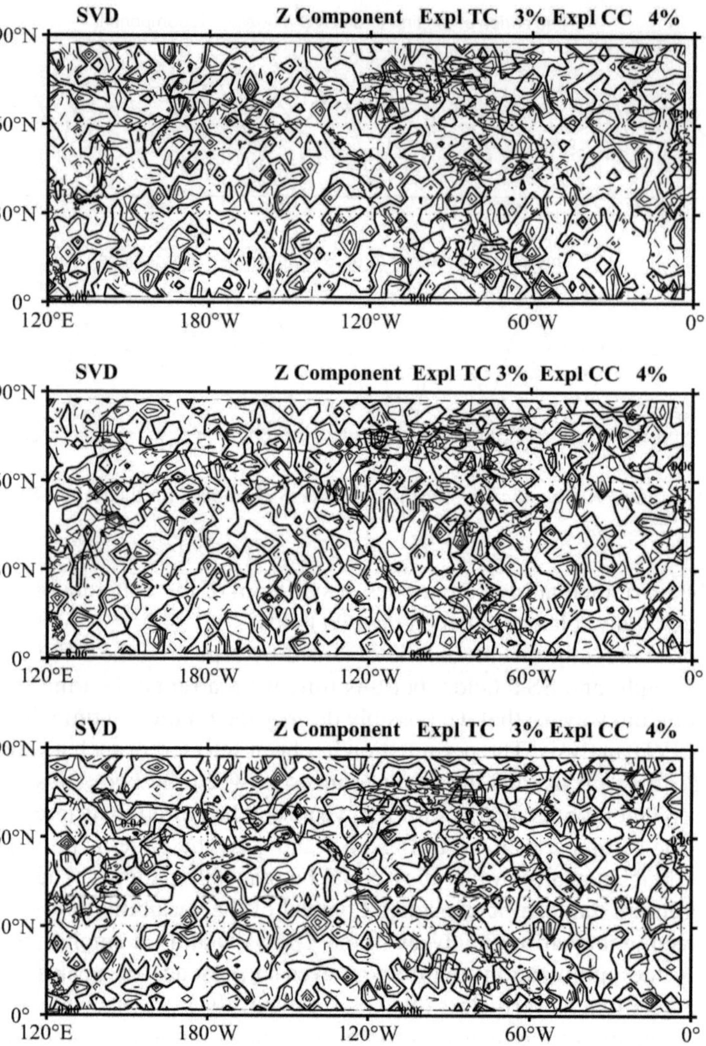

Fig. 6.4 The first three SVD modes for a randomly chosen left and right field data set. Here is shown the **Z** component in descending order of explained total combined variance

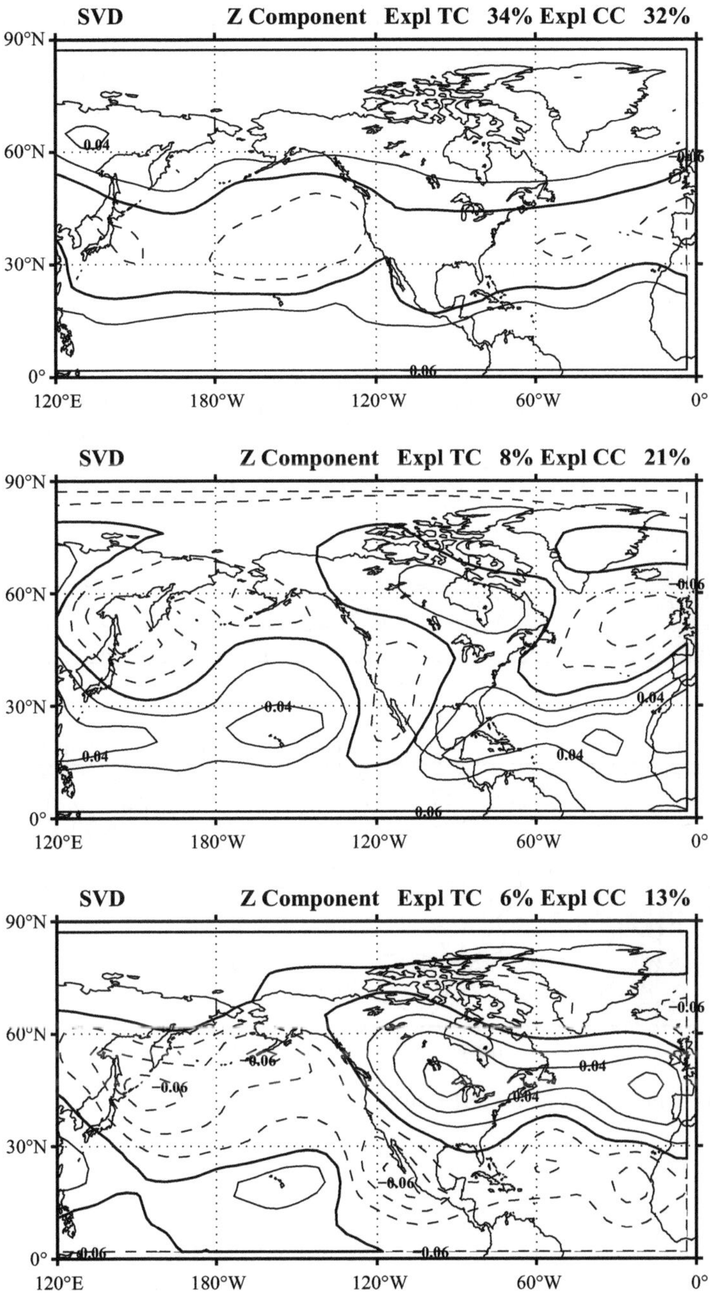

Fig. 6.5 The first three SVD modes for the height and a randomly permuted SST data set. Here is shown the **Z** component in descending order of explained total combined variance

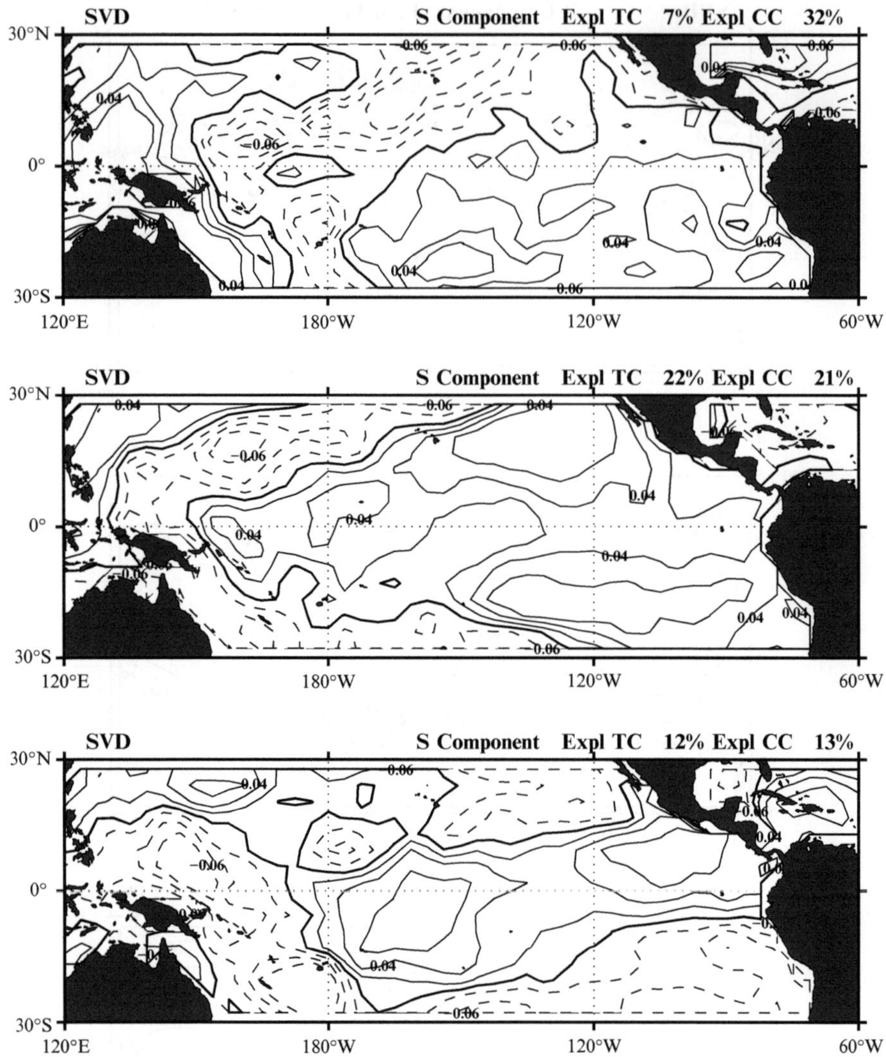

Fig. 6.6 As in Fig. 6.5 but for the SST modes

Chapter 7
The Canonical Correlation Analysis

7.1 The Classical Canonical Correlation Analysis

The method based on the Singular Value Decomposition described in previous chapters was able to represent the largest amount of cross-covariance with fewest modes. The computed modes are designed so as to specifically explain most of the spatial variance. The spatial view point is a requirement that can be put on the modes, but it is hardly unique. Another commonly employed point of view arises if one considers the temporal variations of the modes. In this case one is interested in modes that generate maximally correlated *time* coefficients. This method has been developed in multi-variate analysis and it is known as "Canonical Correlation Analysis". In mathematical terms we seek a decomposition of the data matrices such that the time coefficients are defined as

$$z_k = \sum_{i=1}^{n} u_i a_{ik}, \qquad \text{and} \qquad s_k = \sum_{i=1}^{n} v_i b_{ik},$$

where u, v are unknown patterns and a, b are the time coefficients, so that the entire fields can be reconstructed. We require that the coefficients taken as a time series are uncorrelated with each other,

$$\langle s_k, s_l \rangle = \delta_{kl},$$

where δ_{kl} satisfies $\delta_{kl} = 0$ for $k \neq l$ and $\delta_{kk} = 1$, and the angular bracket denotes the scalar product (or correlation operation). The spatial fields u and v are to be determined as those particular spatial distributions that enforce orthogonality of the time coefficients. Assuming $\mathbf{Z}$ and $\mathbf{S}$ of full row rank, the solution to the problem can be formally obtained by computing an SVD on a normalized version of the cross-covariance matrix. In the following we first reproduce the classical description of the procedure, and then present a computationally more elegant and sound approach that yields a mathematically equivalent solution. We write the SVD on the normalized matrices as

$$(\mathbf{ZZ}^*)^{-\frac{1}{2}}\mathbf{ZS}^*(\mathbf{SS}^*)^{-\frac{1}{2}} = \mathbf{U\Sigma V}^*. \tag{7.1}$$

A. Navarra and V. Simoncini, *A Guide to Empirical Orthogonal Functions for Climate Data Analysis*, DOI 10.1007/978-90-481-3702-2_7,

The left and right singular vectors cannot be used directly to reconstruct the original data matrix, they must be scaled to obtain the so-called "weight vectors" $\widehat{\mathbf{u}}, \widehat{\mathbf{v}}$,

$$\mathbf{W_Z} := (\mathbf{ZZ}^*)^{-\frac{1}{2}}\mathbf{U}, \qquad \mathbf{W_S} := (\mathbf{SS}^*)^{-\frac{1}{2}}\mathbf{V}. \tag{7.2}$$

The weight vectors, that is the columns of $\mathbf{W_Z}$ and $\mathbf{W_S}$, can be used to derive the time coefficients for the reconstruction of the synthesized data set, $a_{ki} = \langle \mathbf{z}_k, \widehat{\mathbf{u}}_i \rangle$, $b_{ki} = \langle \mathbf{s}_k, \widehat{\mathbf{v}}_i \rangle$, or in matrix notation,

$$\mathbf{A} = \mathbf{Z}^*\mathbf{W_Z}, \qquad \mathbf{B} = \mathbf{S}^*\mathbf{W_S}. \tag{7.3}$$

The matrices $\mathbf{A}$ and $\mathbf{B}$ have orthonormal columns (corresponding to time). Indeed,

$$\mathbf{A}^*\mathbf{A} = \mathbf{W}_Z\mathbf{ZZ}^*\mathbf{W}_Z = \mathbf{U}^*(\mathbf{ZZ}^*)^{-\frac{1}{2}}\mathbf{ZZ}^*(\mathbf{ZZ}^*)^{-\frac{1}{2}}\mathbf{U} = \mathbf{I}. \tag{7.4}$$

A similar result holds for $\mathbf{B}$. The cross-product of $\mathbf{A}$ and $\mathbf{B}$ yields an interesting result:

$$\begin{aligned}\mathbf{A}^*\mathbf{B} &= \mathbf{W}_Z^*\mathbf{ZS}^*\mathbf{W}_S \\ &= \mathbf{U}^*(\mathbf{ZZ}^*)^{-\frac{1}{2}}\mathbf{ZS}^*(\mathbf{SS}^*)^{-\frac{1}{2}}\mathbf{V} = \mathbf{U}^*\mathbf{U}\boldsymbol{\Sigma}\mathbf{V}^*\mathbf{V} = \boldsymbol{\Sigma},\end{aligned} \tag{7.5}$$

where we have used (7.1) and the orthogonality property of the left and right singular vectors in the second line. The cross-correlation in time between the coefficients of the same modes is then a measure of the singular values of the cross-covariance matrix scaled as in (7.1).

In practice, the explicit computation of the inverse square roots in (7.1) is avoided by first computing the SVD of the two matrices $\mathbf{Z}$ and $\mathbf{S}$. Let

$$\mathbf{Z} = \mathbf{U_Z}\Sigma_\mathbf{Z}\mathbf{V}_\mathbf{Z}^*, \qquad \mathbf{S} = \mathbf{U_S}\Sigma_\mathbf{S}\mathbf{V}_\mathbf{S}^* \tag{7.6}$$

be the so-called "economy size" SVD of $\mathbf{Z}$ and $\mathbf{S}$, respectively.[1] Therefore,

$$(\mathbf{ZZ}^*)^{-\frac{1}{2}}\mathbf{Z} = \mathbf{U_Z}(\Sigma_\mathbf{Z}\Sigma_\mathbf{Z}^*)^{-\frac{1}{2}}\mathbf{U}_\mathbf{Z}^*\mathbf{U_Z}\Sigma_\mathbf{Z}\mathbf{V}_\mathbf{Z}^* = \mathbf{U_Z}\widehat{\mathbf{V}}_\mathbf{Z}^*,$$

where $\widehat{\mathbf{V}}_\mathbf{Z}$ is the matrix containing the first columns of $\mathbf{V_Z}$, so as to match the size of $\mathbf{U_Z}$. Here we have used the fact that $\Sigma_\mathbf{Z} = [\mathrm{diag}(\sigma_1, \ldots, \sigma_k), O]$. An analogous derivation follows for $\mathbf{S}$. Hence

$$(\mathbf{Z}^*\mathbf{Z})^{-\frac{1}{2}}\mathbf{ZS}^*(\mathbf{SS}^*)^{-\frac{1}{2}} = \mathbf{U_Z}\widehat{\mathbf{V}}_\mathbf{Z}^*\widehat{\mathbf{V}}_\mathbf{S}\mathbf{U}_\mathbf{S}^*.$$

Computing the SVD of this latter matrix, that is

$$\mathbf{U_Z}\widehat{\mathbf{V}}_\mathbf{Z}^*\widehat{\mathbf{V}}_\mathbf{S}\mathbf{U}_\mathbf{S}^* = \mathbf{U}\boldsymbol{\Sigma}\mathbf{V}^*, \quad \text{so that} \quad \widehat{\mathbf{V}}_\mathbf{Z}^*\widehat{\mathbf{V}}_\mathbf{S} = \mathbf{U}_\mathbf{Z}^*\mathbf{U}\boldsymbol{\Sigma}\mathbf{V}^*\mathbf{U_S},$$

[1] The term "economy size" in the factorization $\mathbf{U}\boldsymbol{\Sigma}\mathbf{V}^*$ refers to the fact that only the columns of $\mathbf{U}$ and $\mathbf{V}$ corresponding to nonzero diagonal elements in Σ are retained. Therefore, assuming that the given matrix has maximum rank, either $\mathbf{U}$ or $\mathbf{V}$ is rectangular.

and noticing that $\mathbf{U}_\mathbf{Z}^*\mathbf{U}$ and $\mathbf{V}^*\mathbf{U}_\mathbf{S}$ are orthonormal matrices, reveals that the singular values in Σ are the cosines of the canonical angles between the spaces spanned by $\mathbf{Z}$ and $\mathbf{S}$; see, e.g., Meyer (2000). The smaller the singular values, the farther away the two spaces. Note that the decomposition above yields the same matrices $\mathbf{U}$, Σ, $\mathbf{V}$ as that in (7.1). With these tools, we can easily derive the coefficient matrices $\mathbf{A}$ and $\mathbf{B}$ as follows:

$$\mathbf{A} = \mathbf{Z}^*(\mathbf{Z}^*\mathbf{Z})^{-\frac{1}{2}}\mathbf{U} = \widehat{\mathbf{V}}_\mathbf{Z}\mathbf{U}_\mathbf{Z}^*\mathbf{U}, \quad \mathbf{B} = \widehat{\mathbf{V}}_\mathbf{S}\mathbf{U}_\mathbf{S}^*\mathbf{V}.$$

It readily follows that $\mathbf{A}$ and $\mathbf{B}$ have orthonormal columns, and that $\mathbf{A}^*\mathbf{B} = \Sigma$, as explicitly shown earlier.

The procedure outlined in the previous paragraph performs a correlation analysis (by means of an SVD) with the orthogonal bases of the spaces spanned by $\mathbf{Z}$ and $\mathbf{S}$, allowing to eliminate spurious small correlation singular values which might be due to almost linear dependence among the columns of each of the two matrices $\mathbf{Z}$ and $\mathbf{S}$, and not to a true lack of correlation between the two fields. We will explore this fact more in detail in Sect. 7.3.

The reconstruction of the data sets after a correlation analysis is carried out, is trickier than in previous cases. We recall that $\mathbf{Z}$ and $\mathbf{S}$ have full row rank, from which it follows that the two matrices $\mathbf{W}_\mathbf{Z}$ and $\mathbf{W}_\mathbf{S}$ defined in (7.2) are square and nonsingular, and it holds $(\mathbf{W}_\mathbf{Z})^{-*} = (\mathbf{Z}\mathbf{Z}^*)^{\frac{1}{2}}\mathbf{U}$, $(\mathbf{W}_\mathbf{S})^{-*} = (\mathbf{S}\mathbf{S}^*)^{\frac{1}{2}}\mathbf{V}$. Therefore, from (7.3) it follows

$$\mathbf{Z} = (\mathbf{W}_\mathbf{Z})^{-*}\mathbf{A}^* =: \mathbf{P}_\mathbf{Z}\mathbf{A}^*, \quad \mathbf{S} = (\mathbf{W}_\mathbf{S})^{-*}\mathbf{B}^* =: \mathbf{P}_\mathbf{S}\mathbf{B}^*.$$

We can see that we can reconstruct the data matrix as a combination of the column of the matrix $\mathbf{P}_\mathbf{Z}$ and the coefficients of the matrix $\mathbf{A}$. The rows of $\mathbf{A}$ provide the contribution of each column of $\mathbf{P}_\mathbf{Z}$ to a specific column, that is to say to a specific time, of the data set $\mathbf{Z}$. The columns of $\mathbf{A}$ are the time evolution of the coefficients for a specific column of $\mathbf{P}_\mathbf{Z}$ through the time evolution of the matrix $\mathbf{Z}$. The columns of the matrix $\mathbf{P}_\mathbf{Z}$ are called the "patterns".

The price we are paying to maximize the covariance in the time evolution, is given by this complication of "weight vectors" and "patterns" that coincide in the SVD case, realizing a true basis in the data space. Weight vectors and patterns are not orthogonal. This issue creates problems for the interpretation of the explained variance. We can still reconstruct the data based on the result from the analysis, but in this case the data are a linear combination of the patterns according to the projection of the data on the weight vectors.

7.2 The Modes

Before addressing this interpretation problem, let us have a look at the CCA modes themselves, as they are represented by the patterns $\mathbf{P}_\mathbf{Z}$, $\mathbf{P}_\mathbf{S}$.

The first CCA mode of the height field in Fig. 7.1 is reminiscent of the SVD modes found earlier (Fig. 6.1), but there are significant differences. In the case of

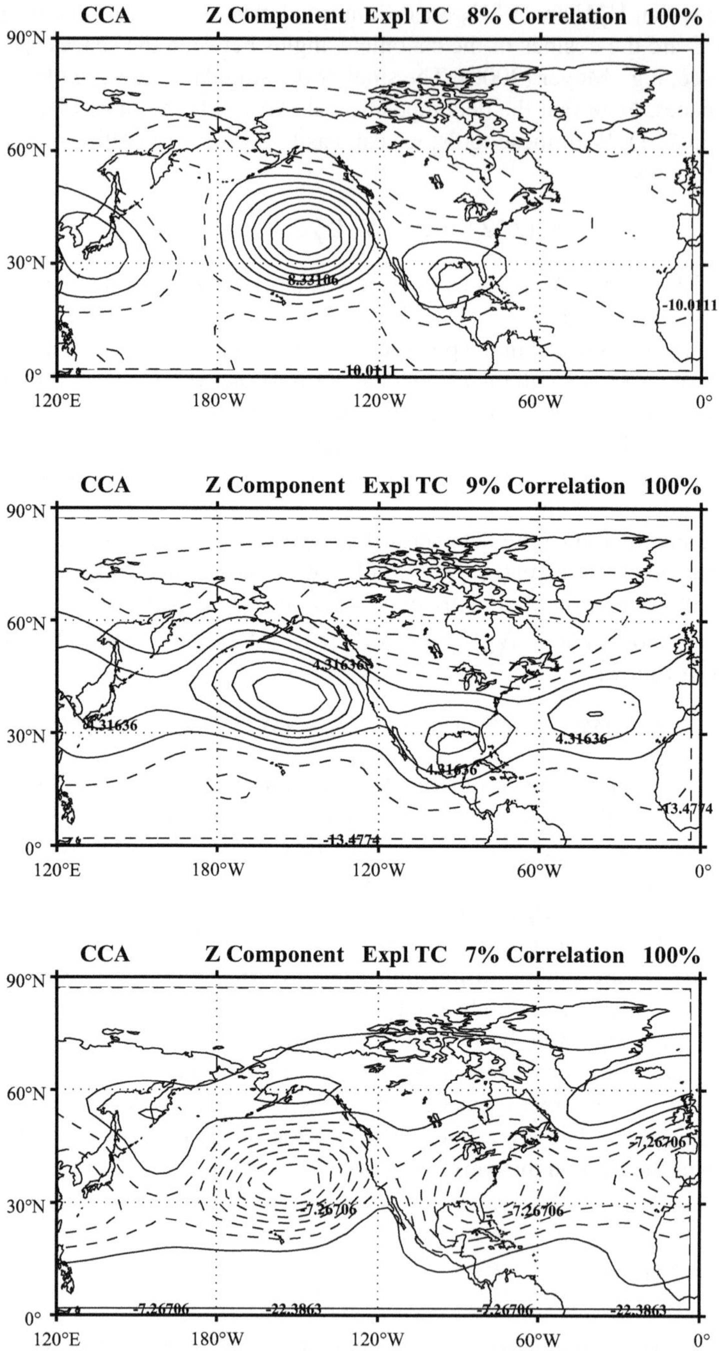

Fig. 7.1 The first three classical CCA modes for the Height-SST data set. Here is shown the **Z** component in descending order of explained total combined variance

the first mode there is a large similarity in shape between the two methods. Both patterns exhibit several lobes over the Central Pacific, the Southern United States and the central Atlantic, whereas a contrasting center of action is over Canada. The intensity contrast is however greater in the case of the CCA and the contrasts between negative and positive centers appear to be smaller in the case of the SVD mode.

There are more differences for the second mode in Fig. 7.1. The SVD mode is somewhat incoherent and it is difficult to assign a particular interpretation.

The second CCA mode shows well defined centers of action over the Central Pacific and the Atlantic ocean, that can be organized in dipoles over the main ocean basin. The dipole can be interpreted as expression of variability associated with jet stream shifting, identified by the strong gradients in the height fields, in a meridional direction. The two dipoles identify the location of two such areas over the main ocean basins.

The SST modes in Figs. 7.2 and 6.2 give a similar interesting picture. The first CCA mode is also similar to the first SVD mode, as in the previous case. The pattern is broad over the Pacific, with the characteristic wedge shape of the variability of the SST in the area. The differences increase in the second mode. The SVD mode tries to capture the main contrast between the east and west portions of the basin, whereas the second CCA mode is concentrated in the equatorial region of the Central Pacific. The CCA mode appears to have a "simpler" structure than the SVD mode and there is no sign of the east–west contrast apparent in the SVD mode.

The third mode is even more striking as the CCA mode attempts to capture the variability of the wedge area margins, as it can be seen by the strong intensification of the mode amplitude at the area border. The SVD mode is a rather complicated creature, trying to focus on some detail of the variability in the West Pacific. Therefore, CCA and SVD "capture different animals in the savana" of the climate variability; how can we argue in favor of one individual or another?

The differences arise mainly from relaxing the orthogonality requirement in the CCA case. A careful observer can visually realize that the CCA modes are not spatially orthogonal. The center of action in the Pacific for the first height CCA mode, for instance, is almost in the same location of the center of action with the same sign in the second mode. The first and second CCA SST modes have also strong overlaps of the same sign in the Central Equatorial Pacific. Clearly those modes are not orthogonal. The CCA method, without the nuisance of generating mutually orthogonal patterns, can find more complicated patterns, as long as they maximize the correlation between the respective time series.

Deciding which mode is more important becomes tricky. The SVD provides a natural ranking of the modes because the orthogonality allows a clean separation of the variance among all the modes. In the CCA case this is no longer possible. We can still have a clue by computing the portion of variance explained by each CCA mode after the data are reconstructed with just that mode, but the overlap between the modes will result in a sum of explained variances that is not equal to the total variance. Indeed, each CCA mode contains a little bit of the others, since the projection of each mode on the other is not zero; this is another way of

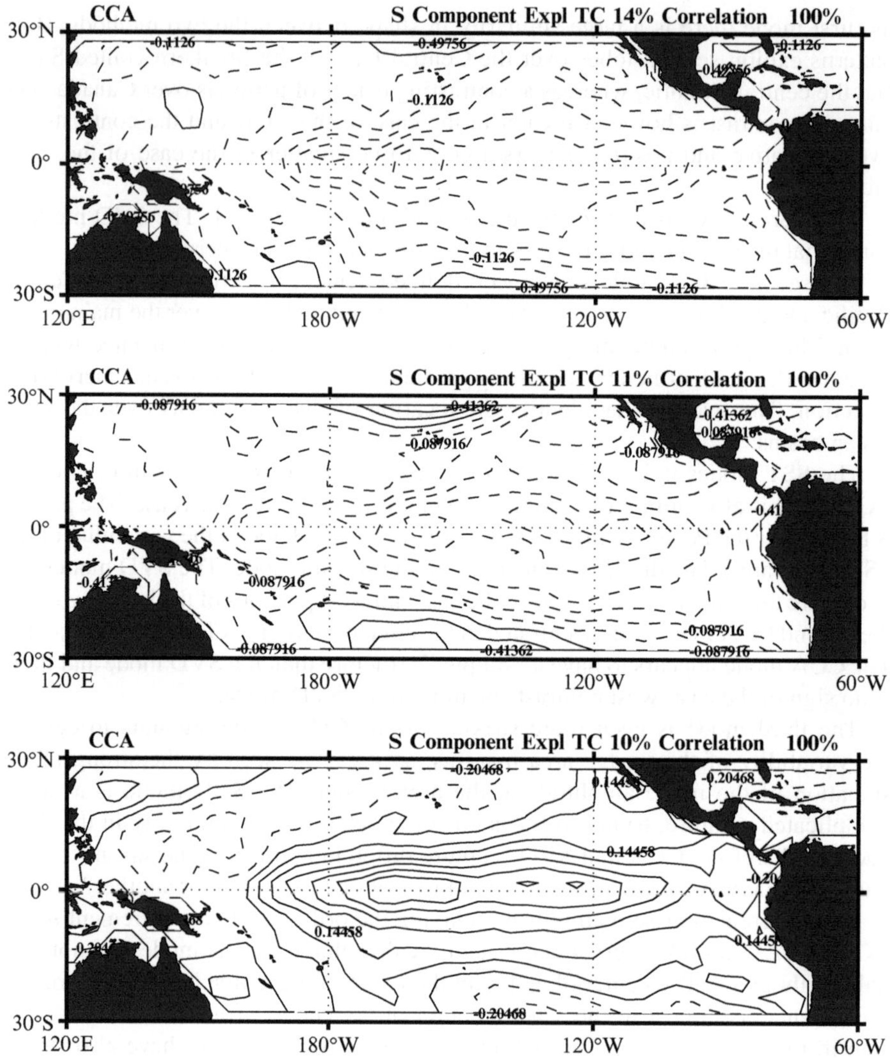

Fig. 7.2 The first three CCA modes for the Height-SST data set. Here is shown the SST component in descending order of explained total combined variance

interpreting their non-orthogonality. One can use the computed explained variance to rank the modes in order of decreasing explained variance, like in the EOF or SVD case. The total variance explained in Figs. 7.1 and 7.2 is computed in such a way. We can see that the result is still acceptable; in practice the explained variances are not completely off the wall because the amount of overlap between modes is often small, but it is clear that the idea of explained variance in the CCA case must be taken with great care.

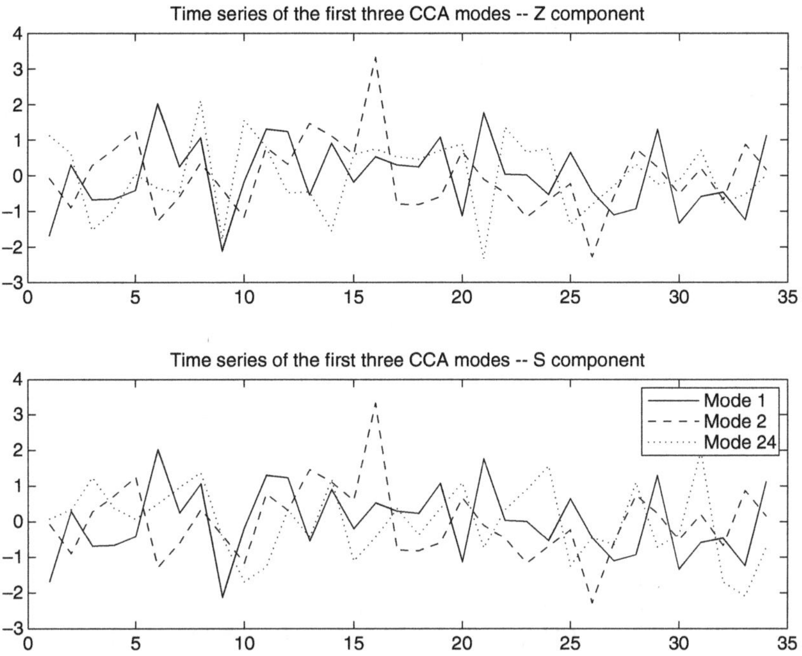

Fig. 7.3 Time series coefficients for the first three CCA modes for the Height-SST data set

The CCA was designed to maximize correlation between time coefficients. Figure 7.3 shows the time series of the time coefficient corresponding to the first two modes in Figs. 7.1 and 7.2. It is possible to see that the time series are strongly correlated. The dotted line, corresponding to mode 24, is less correlated.

This result shows that CCA faces an increasingly difficult task as the modes are found, since the modes must be maximally correlated between height and SST components, but they must be uncorrelated in time with the modes already found. In principle one could use the time correlation to rank the modes, ordering them in decreasing correlation magnitude, but in practice it is easy to find several modes with high values of correlation with a practical degeneracy. In that case we have to go back to the ranking based on the amount of variance explained.

Exercises and Problems

1. Given the two matrices

$$\mathbf{Z} = \begin{pmatrix} 1 & 1 & 1 & 1 & 1 & 1 \\ -1 & 0 & 1 & 2 & 0 & 0 \\ 0 & 1 & 1 & 1 & -2 & 1 \end{pmatrix}, \quad \mathbf{S} = \begin{pmatrix} 1 & -1 & 1 & -1 & 1 & -1 \\ -1 & 1 & 1 & 2 & 0 & 0 \\ -2 & 0 & 2 & 4 & 0 & 0 \end{pmatrix},$$

find the singular values of the scaled cross-covariance in (7.1) and check that if one of the singular values is equal to one then the corresponding time coefficients are equal.

The nonzero singular values are $\Sigma = \text{diag}(1, 0.56363, 0.53860)$. *Explicit computation shows that*

$$
\mathbf{A} = \begin{pmatrix}
-4.0825e-01 & 4.0131e-01 & -4.2683e-01 \\
-5.2830e-16 & 1.5132e-01 & -5.1883e-01 \\
4.0825e-01 & 1.4079e-01 & -2.9290e-01 \\
8.1650e-01 & 1.3026e-01 & -6.6968e-02 \\
8.3039e-17 & 8.6969e-01 & 4.3494e-01 \\
-5.2830e-16 & 1.5132e-01 & -5.1883e-01
\end{pmatrix},
$$

$$
\mathbf{B} = \begin{pmatrix}
-4.0825e-01 & 3.0850e-01 & 8.5980e-02 \\
-1.2630e-15 & 2.6847e-01 & -9.6329e-01 \\
4.0825e-01 & 6.1700e-01 & 1.7196e-01 \\
8.1650e-01 & -1.5425e-01 & -4.2990e-02 \\
-1.1948e-16 & 4.6275e-01 & 1.2897e-01 \\
1.1948e-16 & -4.6275e-01 & -1.2897e-01
\end{pmatrix},
$$

confirming that the first column is indeed the same, up to machine precision.

2. Each CCA mode for height and SST can be considered a vector in the data space. Compute the angles between the CCA modes for height and SST.

7.3 The Barnett–Preisendorfer Canonical Correlation Analysis

As already mentioned, the explicit computation of $(\mathbf{ZZ}^*)^{-\frac{1}{2}}, (\mathbf{SS}^*)^{-\frac{1}{2}}$ in the classical CCA is prone to instability problems, in case the matrices $\mathbf{ZZ}^*$ and $\mathbf{SS}^*$ are singular or almost so, that is, in case $\mathbf{Z}$ and $\mathbf{S}$ are not numerically full (row) rank matrices. Mathematically speaking, if the smallest eigenvalue of $\mathbf{ZZ}^*$ is different from zero by tiny, say 10^{-15}, it is enough to declare the matrix invertible; however, this is not sufficient from a *numerical* standpoint. The inverse matrix will be dominated by the smallest eigenvalue (see Sect. 2.9) that will contribute predominantly to the structure of the inverse. We could neglect that, were it not for the fact that the smallest eigenvalues are the most sensitive to perturbations of the original matrix, i.e. the covariance matrix. Such perturbations can be variations in the number of columns that represent a change in the number of time levels represented in the data, namely a change in the time sampling. Variations in the rows, implying changes in the spatial domain or in the space sampling, can be just as damaging. The inverse square root function is very sensitive to small changes in the composition of the data matrix, introduced for instance by random errors in the data. The effect can show up in various ways, for instance as spurious large correlations. We have already encountered this problem in Chap. 4 and we have seen that the EOF can be very effective in filtering out the noise present in the data. Barnett and Preisendorfer in 1987 proposed to use EOF to filter the data before applying CCA in order to minimize the unduly relevance of the random errors in the calculation of the scaling matrices; see Barnett et al. (1987). More generally, this strategy may be used to eliminate spurious (or noisy) information from the original data matrices. In the following we report the procedure to obtain the modes as described by Barnett and Preisendorfer (Fig. 7.4–7.5). It should be clear to the reader that the whole process may be derived by first truncating the SVD of the two matrices $\mathbf{Z}$, $\mathbf{S}$ in (7.6) to the first dominant singular values.

The approach follows the lines of the classical CCA analysis, but the original covariance matrices are now expanded preliminarily on the respective EOF,

$$\mathbf{ZZ}^* = \mathbf{E}_Z \boldsymbol{\Sigma}_Z \mathbf{E}_Z^*, \qquad \mathbf{SS}^* = \mathbf{E}_S \boldsymbol{\Sigma}_S \mathbf{E}_S^*, \tag{7.7}$$

where the $\mathbf{E}$ represent the EOF for $\mathbf{Z}$ and $\mathbf{S}$ respectively and $\boldsymbol{\Sigma}$ is the diagonal matrix of the variances. The data is then projected onto the respective EOF basis to obtain primed quantities

$$\mathbf{Z}' = \mathbf{E}_Z^* \mathbf{Z}, \qquad \mathbf{S}' = \mathbf{E}_S^* \mathbf{S}. \tag{7.8}$$

The covariance matrix expressed in terms of the primed quantities, the EOF coefficients, is particularly simple

$$\mathbf{Z}'\mathbf{Z}'^* = \mathbf{E}_Z^* \mathbf{ZZ}^* \mathbf{E}_Z^* = \mathbf{E}_Z^* \mathbf{E}_Z \boldsymbol{\Sigma}_Z \mathbf{E}_Z^* \mathbf{E}_Z = \boldsymbol{\Sigma}_Z, \tag{7.9}$$

with a similar expression for the covariance matrix for $\mathbf{S}'$. In the Barnett–Preisendorfer (BP) approach, the EOF coefficients are normalized and then a certain number of them is discarded, retaining only a smaller number. This operation corresponds to removing the higher order EOF that, as we have seen, are the most affected by noise and sampling errors. In practice this operation is performed by removing selected rows in the matrices $\mathbf{Z}$ and $\mathbf{S}$, to obtain reduced matrices of EOF coefficients, indicated by the tilde sign,

$$\mathbf{X} = \begin{bmatrix} x_1(1) & x_1(2) & \dots & x_1(n) \\ x_2(1) & x_2(2) & \dots & x_2(n) \\ x_3(1) & x_3(2) & \dots & x_3(n) \\ \dots & \dots & \dots & \dots \\ x_m(1) & x_m(2) & \dots & x_m(n) \end{bmatrix} \rightarrow \textit{Reduce} \rightarrow \begin{bmatrix} x_1(1) & \dots & x_1(n) \\ x_2(1) & \dots & x_2(n) \end{bmatrix} = \tilde{\mathbf{Z}}.$$

The number of retained modes may be different for the left and right fields, because the BP-CCA method requires performing an SVD of the cross-covariance matrix of the normalized reduced coefficients matrix (tilde) quantities:

$$\widetilde{\mathbf{ZS}^*} = \mathbf{U}\boldsymbol{\Sigma}\mathbf{V}^*. \tag{7.10}$$

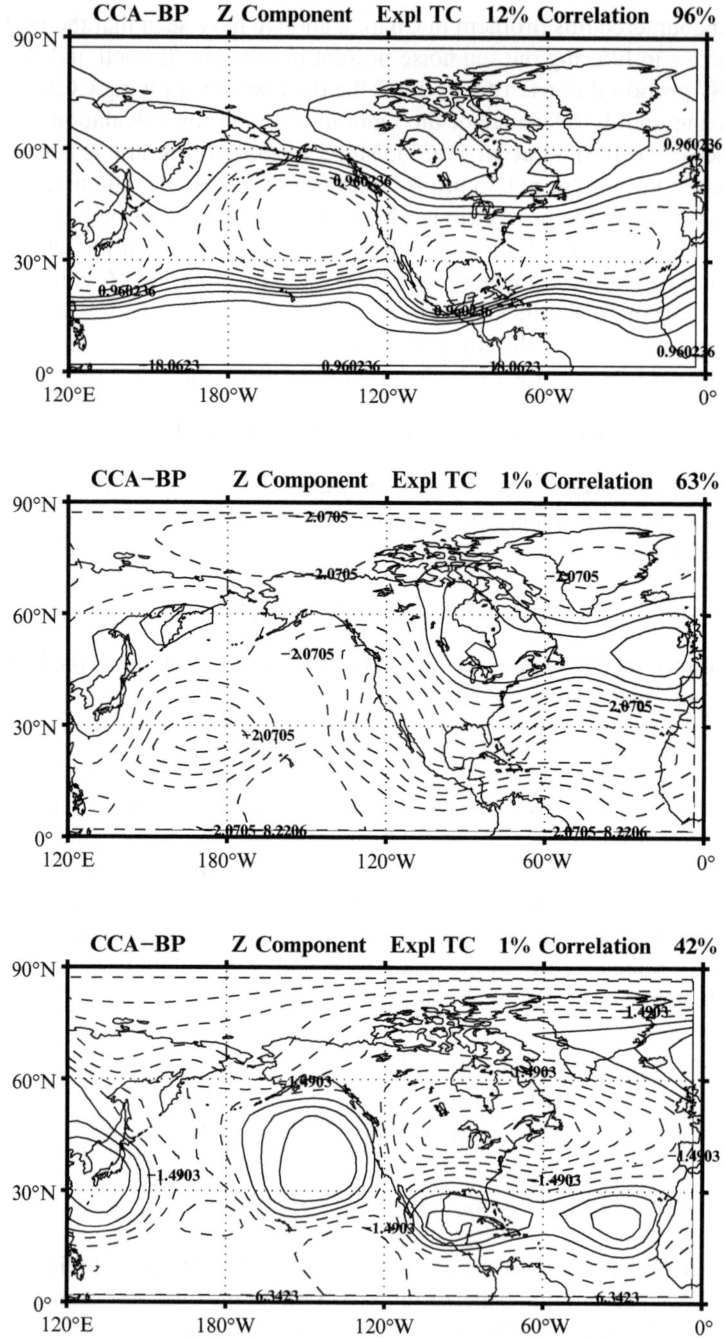

Fig. 7.4 The first three Barnett–Preisendorfer CCA modes for the Height-SST data set. Here is shown the **Z** component in descending order of canonical correlations

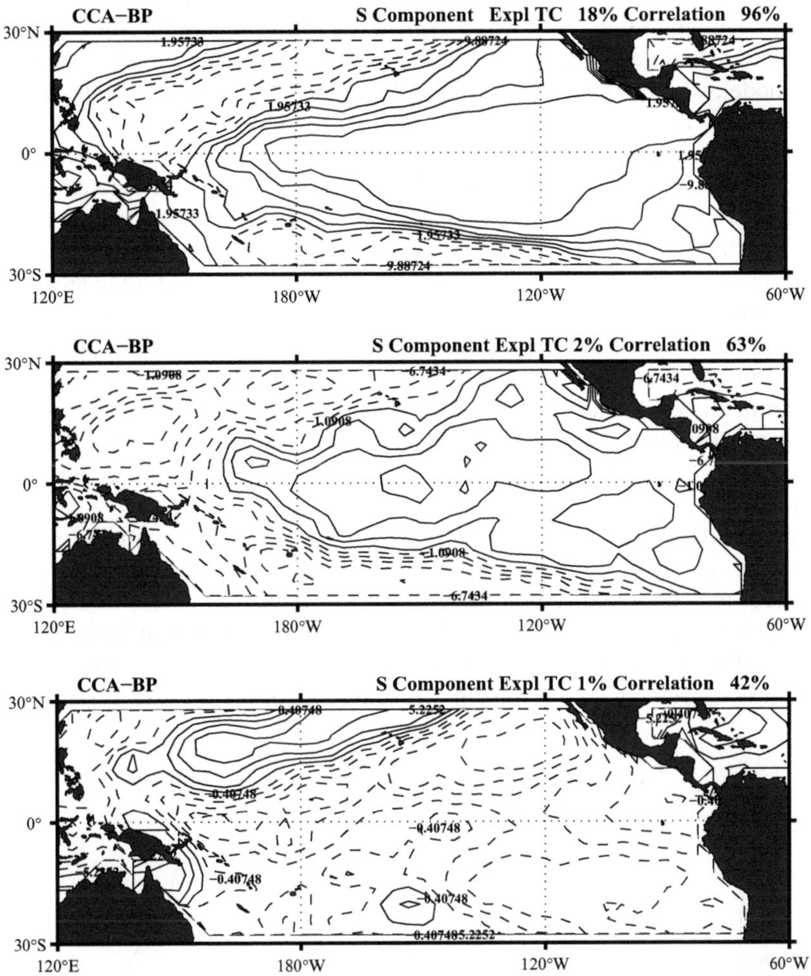

Fig. 7.5 The first three CCA Barnett–Preisendorfer modes for the Height-SST data set. Here is shown the SST component in descending canonical correlation order

This only requires that the number of time levels (the number of columns) be the same for the left and right fields. The "weight vectors" $\widehat{\mathbf{u}}, \widehat{\mathbf{v}}$ are then calculated as before, exploiting the simplified version of the covariance in (7.9). The data can then be reconstructed using the coefficients computed with the weight vectors,

$$(\boldsymbol{\Sigma}_Z)^{-\frac{1}{2}}\mathbf{U} = \mathbf{W}_Z, \qquad (\boldsymbol{\Sigma}_S)^{-\frac{1}{2}}\mathbf{V} = \mathbf{W}_S, \tag{7.11}$$

and $\mathbf{A} = \tilde{\mathbf{Z}}^*\mathbf{W}_Z$, $\mathbf{B} = \tilde{\mathbf{S}}^*\mathbf{W}_S$, where, as before, the matrices $\mathbf{A}$ and $\mathbf{B}$ hold the coefficients required to reconstruct the data.

The first modes are not too different from the classical CCA and we can see that we can recover the familiar distribution of the centers of action in the height field

(Fig. 7.5) and the wedge shape that is typical of the SST distribution (Fig. 7.5). The first mode is dominated by the Pacific North American pattern as it is represented in the model, whereas the second mode is highly localized in the Atlantic, pointing to a variability in the position of the Atlantic jet stream. Interpretation of the result is always difficult, especially since these modes express correlations between the height fields and the tropical Pacific SST. A straightforward interpretation has to consider the height modes as those patterns that are most correlated with the corresponding SST pattern. It is probably easy to come to terms with the first **Z** mode, indicating an effect of the tropical SST on the immediate neighboring areas of the north Pacific and North America. Some more sophisticated dynamics are involved to explain the effect of the tropical SST on the Atlantic region where the second **Z** mode is mostly concentrated. Remote effects of SST are well known, but the evidence from this CCA analysis is usually not enough to guarantee such a conclusion. The time series reflects the fact that the correlations are less saturated than in the classical CCA case. The canonical correlations, i.e. the singular values of the cross-covariance matrix, are not clustered around unity and so we can see a drop in the correlation values from the first to the third mode (Fig. 7.6). The time series of the coefficients can be quite different from the classical CCA case, even if the patterns themselves may not seem very different.

There are no rules to recommend the number of EOF to retain in the BP-CCA approach. The spirit here is to retain only the portion of the variance in the original field that is less affected by errors, in this sense any of the comments described in

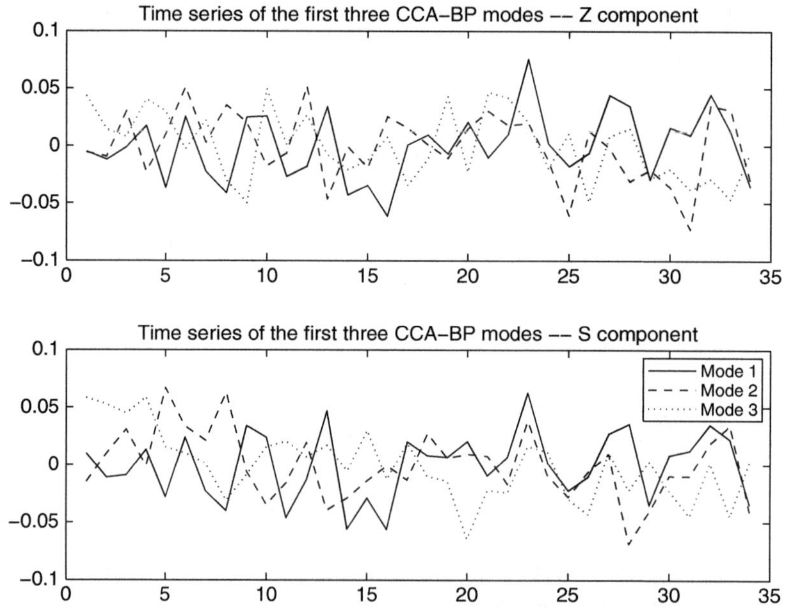

Fig. 7.6 The time series coefficients for the first three CCA Barnett–Preisendorfer modes for the Height-SST data set

Chap. 4 can be used to define a criterion to choose a number of EOF to keep. The sensitivity of the BP-CCA method is however smaller (Fig. 7.6). We can see in Figs. 7.7 and 7.8 that the first mode is not much affected by the changes in the amount of variance retained in the BP-CCA analysis. The second mode is more sensitive, showing marked differences according to the number of EOF kept in the original data sets.

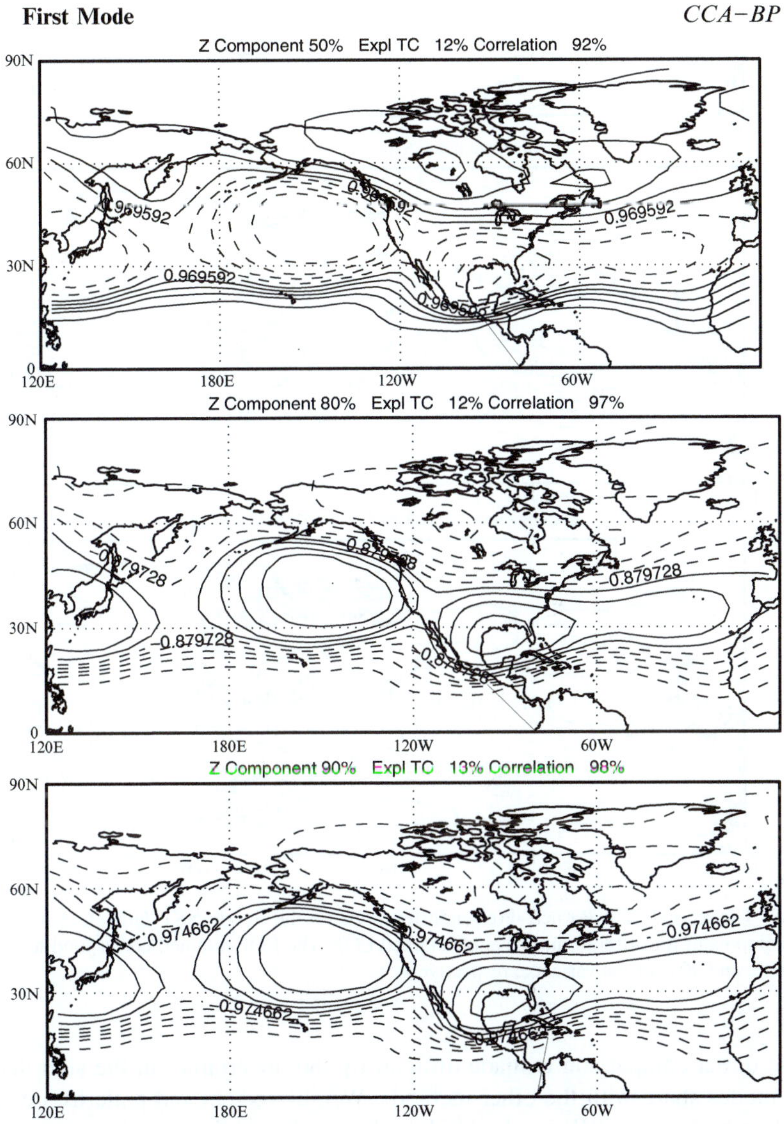

Fig. 7.7 The first CCA Barnett–Preisendorfer mode for the Height-SST data set. Here is shown the **Z** component for different numbers of retained EOF. The EOF retained correspond to keeping 50%, 80% and 90% of the variance, respectively

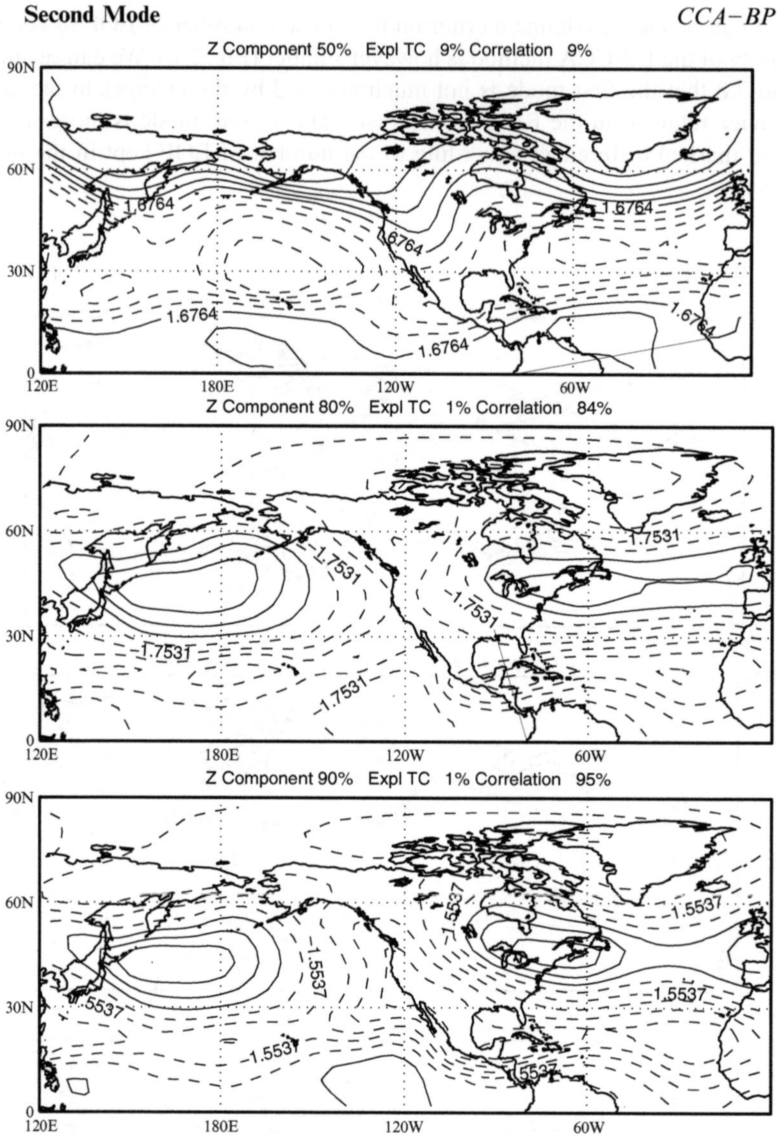

Fig. 7.8 The second CCA Barnett–Preisendorfer mode for the Height-SST data set. Here is shown the **Z** component for different numbers of retained EOF. The EOF retained correspond to keeping 50%, 80% and 90% of the variance, respectively

The direct comparison is made difficult by the uncertainty in the sign that the CCA modes share with the other methods. Weight vectors and patterns are determined up to a an overall sign by CCA, either in the original version or according to the BP prescription. This means that the algorithm used to compute the modes randomly picks one sign. General consistency is guaranteed by the fact that the coefficients sign is changed accordingly. This is easy to recognize in the patterns, but it may become tricky if time series from different calculation are compared. Care must be taken to make sure that the modes are comparable in terms of having all the same sign convention. The strategies used to choose an appropriate sign for the EOF (see Chap. 4) can also be employed here.

Chapter 8
Multiple Linear Regression Methods

8.1 Introduction

In the previous chapters we have introduced techniques to find relations between two or more different fields. In this chapter we describe a more general framework that may provide additional insight into previously analyzed methods. In general a relation between fields can be formulated symbolically as

$$\mathbf{Z} = f(\mathbf{S}) \tag{8.1}$$

where, for instance, **Z** represents the geopotential and **S** the SST. The exact form of the relation is unknown, but it is probably time dependent and thus includes effects of time lags and so on. In practice, it is really difficult to investigate arbitrary functional forms for f in (8.1); assuming f to be linear represents a simplifying but viable alternative. In this case the function f may be represented by matrices. We have seen in the previous chapters that powerful methods have been devised to identify relations of the form (8.1) assuming that $f(\mathbf{S})$ is a linear function. We have seen linear correlation methods, teleconnection analysis and finally methods that analyze systematically the linear relation between two data sets, such as the Singular Value Decomposition (SVD) or Canonical Correlation Analysis (CCA). We will now define a general framework that includes the latter as special case.

Assuming linearity, the relation between the data matrices **Z** and **S** can be written as (Navarra and Tribbia 2005)

$$\mathbf{Z} = \mathbf{AS} \tag{8.2}$$

where **A** is a matrix, assumed here to be stationary. **Z** and **S** are data matrices describing the atmospheric and oceanic fields arranged at fixed times. In general, the number of spatial points need not be the same in the **Z** and **S** data and so the two matrices will be in general rectangular matrices, with n time columns and, respectively, p rows for **Z** and q rows for **S**.

It is possible to set a simple (least squares) minimization problem for **A** (Golub and Van Loan 1996; Richman and Vermette 1993), as

A. Navarra and V. Simoncini, *A Guide to Empirical Orthogonal Functions for Climate Data Analysis*, DOI 10.1007/978-90-481-3702-2_8,

$$\min_{\mathbf{A}} \|\mathbf{Z} - \mathbf{AS}\|_F \tag{8.3}$$

where the norm is the Frobenius norm, as defined in (2.5). If $\mathbf{SS}^T$ is full rank, the minimizing least squares solution can be written as

$$\mathbf{A} = \mathbf{ZS}^T(\mathbf{SS}^T)^{-1}. \tag{8.4}$$

If $\mathbf{S}$ is square, the solution is exact and the residual $\mathbf{Z}-\mathbf{AS}$ is zero. Otherwise the obtained solution minimizes the residual among all possible choices of $\mathbf{A}$, yielding a nonzero residual. If $\mathbf{SS}^T$ is singular, then a solution can still be found by using the pseudoinverse defined in (2.12).

The matrix $\mathbf{A}$ in (8.4) describes the relation between the fields $\mathbf{S}$ and $\mathbf{Z}$, and it is the operation that transforms the $\mathbf{S}$ field into $\mathbf{Z}$.

In the case that the minimum is zero, then the entire field $\mathbf{Z}$ can be transformed into $\mathbf{S}$. In this case $\mathbf{AS} = \mathbf{Z}$ and the range of $\mathbf{A}$ spans the whole space $\mathbf{Z}$. In addition, the variances of $\mathbf{AS}$ and of $\mathbf{Z}$ coincide, namely

$$\text{diag}((\mathbf{AS})(\mathbf{AS})^T) = \text{diag}(\mathbf{ZZ}^T),$$

where we recall that we are still assuming that the fields have zero mean.

If the minimum is not zero, then $\mathbf{AS}$ does not coincide with the space $\mathbf{Z}$ and in general the range of $\mathbf{A}$ will be a subspace of $\mathbf{Z}$. Now only a portion of the field $\mathbf{Z}$'s variability can be associated with the variability of $\mathbf{AS}$. There is a difference between the field $\mathbf{Z}$ and $\mathbf{AS}$ so that

$$\mathbf{Z}_{free} = \mathbf{Z} - \mathbf{AS}.$$

In general we can thus write the field $\mathbf{Z}$ as

$$\mathbf{Z} = \mathbf{AS} + \mathbf{Z}_{free} = \mathbf{Z}_{\mathbf{S}} + \mathbf{Z}_{free}. \tag{8.5}$$

This splitting of $\mathbf{Z}$ makes explicit that a portion of the field $\mathbf{Z}$ can be reached directly from $\mathbf{S}$ via the operator $\mathbf{A}$ (denoted by $\mathbf{Z}_{\mathbf{S}}$), but a residual part, $\mathbf{Z}_{free}$, cannot be reached. It is interesting to observe that the two parts are uncorrelated in time, that is

$$\mathbf{Z}_{\mathbf{S}}\mathbf{Z}_{free}{}^T = 0.$$

Indeed,

$$\begin{aligned}\mathbf{Z}_{\mathbf{S}}\mathbf{Z}_{free}{}^T &= \mathbf{Z}_{\mathbf{S}}(\mathbf{Z}^T - \mathbf{S}^T\mathbf{A}^T) = \mathbf{ASZ}^T - \mathbf{ASS}^T\mathbf{A}^T \\ &= \mathbf{ASZ}^T - \mathbf{ASS}^T(\mathbf{SS}^T)^{-1}\mathbf{SZ}^T = 0.\end{aligned}$$

Exercises and Problems

1. Given the matrices $\mathbf{S} = [1,\ -1,\ 1,\ -1;\ 1,\ 1,\ 1,\ 1]$; and $\mathbf{Z} = [1,\ -1,\ 2,\ 0;\ 0,\ 1,\ -1,\ 2;\ 0,\ 0,\ 1,\ 1]$, compute the least squares solution of the problem $\min_{\mathbf{A}} \|\mathbf{Z} - \mathbf{AS}\|_F$.
   We have

$$A = (ZS^T)(SS^T)^{-1} = \begin{bmatrix} 4 & 2 \\ -4 & 2 \\ 0 & 2 \end{bmatrix} \left(\begin{bmatrix} 4 & 0 \\ 0 & 4 \end{bmatrix} \right)^{-1} = \begin{bmatrix} 1 & 0.5 \\ -1 & 0.5 \\ 0 & 0.5 \end{bmatrix}.$$

2. With the data of the previous exercise, verify that the residual is orthogonal to **AS**.
   A direct computation shows that

$$(AS)Z_{free}^T = \frac{1}{2} \begin{bmatrix} 3 & -1 & 3 & -1 \\ -1 & 3 & -1 & 3 \\ 1 & 1 & 1 & 1 \end{bmatrix} \cdot \frac{1}{2} \begin{bmatrix} -1 & 1 & -1 \\ -1 & -1 & -1 \\ 1 & -1 & 1 \\ 1 & 1 & 1 \end{bmatrix} = 0.$$

8.1.1 A Slight Digression

The search for a relation between the fields can also be formulated in a different way, via the following minimization problem:

$$\min_{\mathbf{B}} \|\mathbf{S} - \mathbf{BZ}\|_F ,$$

where we are trying to get **S** in terms of **Z**. This problem has an analogous minimization solution

$$\mathbf{B} = \mathbf{SZ}^T(\mathbf{ZZ}^T)^{-1}. \tag{8.6}$$

We now have two operators, **A** and **B**, that express the relation between **S** and **Z**, but that are not completely equivalent. The operator **A** can be interpreted as expressing the influence of **S** in terms of **Z**, whereas the operator **B** represents the influence of **Z** on **S**. They both involve the cross-correlation matrix $\mathbf{ZS}^T$ and its transpose $\mathbf{SZ}^T$. In general there is no reason to expect them to have any special structure.

Further insight can be gained by realizing that each of (8.4) and (8.6) represents a multivariate regression problem for the atmospheric field **Z** on the oceanic SST field **S**. This is a general formulation of the coupling problem, that is the identification of the relation between two varying fields. It is only subjected to the linearity constraint in the coupling, and it will give an indication of the strength of the relation between one field and the other. The method based on the matrices **A** and **B** obtained via the Least Squares method will be denoted in the following as the PRO method. This is reminiscent of the name, *PROcrustes* problem, commonly employed for this formulation in the climatology community, although the true Procrustes problem requires additional constraints, and it is discussed in Sect. 8.2.2.

It is interesting to note that the method based on the pseudoinverse can be applied to any pair of fields. We also have made no assumptions regarding the geographical location of the data we are using for the analysis. The fields **S** and **Z** could be located in the same geographical domain or they could be placed in remote locations distant from each other. We might consider, for instance, the geopotential and the SST in the same domain in the tropics, or we might take the tropical SST and the geopotential over North America. In the former case we are looking at local relation between the fields, in the latter case we are really looking at remote influences, probably mediated by other physical processes.

8.2 A Practical PRO Method

The cost of the calculation described in the preceding section depends strongly on the order of the data matrices. It is a function of the row and column dimension of **Z** and **S**. The cross-correlation matrix $\mathbf{Z}\mathbf{S}^T$, that is the essential part of **A**, may have very large dimension if many grid points are considered. In our case, its dimensions are $p \times q$. In some applications it is not a problem, but for a typical climate or meteorological application the number of grid points can quickly run into the thousand, making the calculation of $\mathbf{Z}\mathbf{S}^T$ unpractical.

A significant simplification of the calculation can be achieved by using the data compression properties of the EOF. Using the EOF we have introduced in previous chapters we can achieve a significant reduction in the problem size to be solved. The maximum number of EOF for a data field, say **Z**, is given by the smaller dimension of **Z**. In typical meteorological applications the number of time levels is often much smaller than the number of grid points and we can reduce the problem significantly.

If the columns of $\mathbf{U_Z}$, $\mathbf{U_S}$ contain the EOFs of the two fields **Z**, **S**, respectively, that is, their left singular vectors, then we know from Chap. 4 that

$$\mathbf{Z} = \mathbf{U_Z}\tilde{\mathbf{Z}}, \quad \mathbf{S} = \mathbf{U_S}\tilde{\mathbf{S}}.$$

We can then define the significantly smaller problem in terms of the EOF coefficients as

$$\min_{\mathbf{A}} \|\tilde{\mathbf{Z}} - \mathbf{A}\tilde{\mathbf{S}}\|_F.$$

Its solution can be found in a similar way in terms of the tilde quantities as

$$\mathbf{A} = \tilde{\mathbf{Z}}\tilde{\mathbf{S}}^T(\tilde{\mathbf{S}}\tilde{\mathbf{S}}^T)^{-1}. \tag{8.7}$$

The reduction of the algebraic dimension of the problem is quite significant. In the case of a geophysical field, the data matrices are usually very rectangular, because the number of columns describing the spatial extent of the field is usually much larger than the number of rows describing the number of time levels analyzed. The minimization problem is then quite tractable. The use of EOFs also offers the possibility of an interpretation of the operators **A** and **B**. The operator **A** expresses the contribution to a single mode of **Z**, for instance the first mode by all the modes of **S**. By inspecting the columns of **A** we can analyze the regression factor by which each mode of **S** contributes to that particular mode. Large values indicate a strong impact of that **S** mode on the variability of the first **Z** mode. The analysis can be repeated for each column, thereby reconstructing the map of the **S** modes that have strong influences on **Z**. A similar argument can be done for the operator **B**, in which the role of **S** and **Z** are reversed. In this case the column will indicate which of the **Z** modes contributed more strongly to the first **S** mode. Together, the two operators contain a fairly detailed map of the influence patterns between the fields.

The operator **B** can be interpreted in a similar way, with the role of **Z** and **S** reversed. Now the (1,1) component of **B** expresses the influence of the first **Z** mode on the first **S** mode or of the second mode on the first mode, and so on.

In principle, the usage of the EOF allows one to filter the data prior to the application of the PRO method, by retaining only some of the EOF and achieving another significant saving. This is not required by the method itself, but is a feature that adds further flexibility to the method and can be helpful in avoiding overfitting.

8.2.1 A Different Scaling

It is interesting to note what happens when the data are scaled by the covariance matrices. If we take the data in the EOF representation and we scale them by the square root of their covariance matrices,

$$\hat{\mathbf{Z}} = (\tilde{\mathbf{Z}}\tilde{\mathbf{Z}}^T)^{-\frac{1}{2}}\tilde{\mathbf{Z}}, \quad \hat{\mathbf{S}} = (\tilde{\mathbf{S}}\tilde{\mathbf{S}}^T)^{-\frac{1}{2}}\tilde{\mathbf{S}}$$

then

$$\hat{\mathbf{Z}}\hat{\mathbf{Z}}^T = \mathbf{I}, \quad \hat{\mathbf{S}}\hat{\mathbf{S}}^T = \mathbf{I}.$$

When this scaling is used the cross-covariance matrix becomes the cross-correlation matrix. The minimizing solution to the scaled least squares problem $\min_{\mathbf{A}} \|\hat{\mathbf{Z}} - \mathbf{A}\hat{\mathbf{S}}\|_F$ can then be written as

$$\mathbf{A} = \hat{\mathbf{Z}}\hat{\mathbf{S}}^T.$$

Interestingly the sister problem can be solved as

$$\mathbf{B} = \hat{\mathbf{S}}\hat{\mathbf{Z}}^T$$

and so $\mathbf{A} = \mathbf{B}^T$.

In this scaling the influence matrices **A** and **B** are one the transpose of the other. This means that only one matrix is sufficient to describe the interaction among the various modes. In this case, the upper half of the matrix describes the influence of **Z** on **S** and the lower half the influence of **S** on **Z**. This scaling is used in the Canonical Correlation Analysis approach and this relation prompts us to examine what are the connections between the PRO methods and the other methods used to analyze variance.

8.2.2 The Relation Between the PRO Method and Other Methods

It is interesting to analyze what happens if further restrictions are put on the coupling matrix **A**. If we require that **A** be an orthonormal matrix **Q**, i.e. $\mathbf{Q}\mathbf{Q}^T = \mathbf{Q}^T\mathbf{Q} = 1$, then we obtain the *orthogonal Procrustes problem*,[1]

$$\min_{\mathbf{Q}} \|\mathbf{Z} - \mathbf{Q}\mathbf{S}\|_F \,, \tag{8.8}$$

whose solution is given by (see, e.g., Golub and Van Loan 1996, sec.12.4.1)

$$\mathbf{Q} = \mathbf{U}\mathbf{V}^T,$$

where **U** and **V** are obtained by the Singular Value Decomposition of the cross-correlation matrix, that is

$$\mathbf{S}\mathbf{Z}^T = \mathbf{U}\Sigma\mathbf{V}^T.$$

This is the definition of the SVD method as proposed by Bretherton et al. (1992), and we can now see that it is essentially a Procrustes problem. This result is consistent with Cherry (1996, 1997) who found that the SVD method essentially aims at rotating one data set into the other. Searching coupled modes with SVD is therefore equivalent to assuming a priori that the coupling relation between the fields is special. Similarly, it is also possible to realize that the Canonical Correlation Analysis imposes a similar orthogonality requirement on **Q**. From this point of view it is

[1] The name is taken from the Greek mythology. Procrustes, the owner of a tavern, had only one bed and therefore took to sawing off the legs of his guests if they were too long for his bed. In a similar way, we are trying to "constrain" the matrix **Z** into **S** and we are willing to chop off some part of **Z** in order to so.

not surprising that identification of coupled modes via SVD or CCA is sometimes arduous, since the orthogonality constraint for the influence operators does not seem to have any physical justification.

8.3 The Forced Manifold

It is often the case in meteorology or climatology that numerical simulations are repeated with similar forcing conditions and slightly different initial conditions. The reason can be found in the extremely sensitive nature of the atmospheric systems to small perturbations in the initial values. Small initial differences can quickly evolve in large differences because of the natural growth of instabilities and other nonlinear feedbacks. This phenomenon makes sometimes difficult the detection of signal imposed on the climate systems by external factors, like for instance a certain prescribed distribution of Sea Surface Temperatures (SST). In the preceding chapters we have often used data from simulations that were derived exactly in that manner, the objective to isolate the effect on the atmospheric variability of the changes in SST. This kind of experiments is often designed as an *ensemble* experiment in which the same SST distribution changing in time month after month is used and several simulations with slightly different initial values are used. A number of statistical methods can then be used on the resulting ensembles to detect the effect of SST.

We can use the PRO method described in the previous sections to extract the signal from these experiments. The data matrix for $\mathbf{Z}$ must be extended to include all the members of the ensemble

$$\mathbf{Z} = [\mathbf{z}_1^a, \mathbf{z}_2^a, \ldots, \mathbf{z}_n^a, \mathbf{z}_1^b, \mathbf{z}_2^b, \ldots, \mathbf{z}_n^b, \ldots],$$

where the superscripts $a, b, \ldots$ label the individual members of the ensemble. The data matrix for $\mathbf{S}$ is obtained by repeating the time series to match the number of members

$$\mathbf{S} = [\mathbf{s}_1, \mathbf{s}_2, \ldots, \mathbf{s}_n, \mathbf{s}_1, \mathbf{s}_2, \ldots, \mathbf{s}_n, \ldots].$$

The PRO method can then be applied to the data matrices $\mathbf{Z}$ and $\mathbf{S}$. Figure 8.1 shows the result of the PRO method when it is applied to the Pacific North American region for the atmospheric field $\mathbf{Z}$ and to the tropical region for the SST. The PRO method divides the field $\mathbf{Z}$ in two orthogonal parts, the first component has maximum correlation with the other field, in this case the SST, the second is uncorrelated from the SST. In mathematical terms the two parts are subspaces of the original data, we can call them the *Forced Manifold* in the first case, to represent the fact that the subspace contains the effects of the forcing field and we may call the other *Free Manifold* to represent its independence of the variations of the forcing field. The figure shows the ratio between the variance of the Forced Manifold and the total variance locally point by point. The orthogonality of the two subspaces can be seen

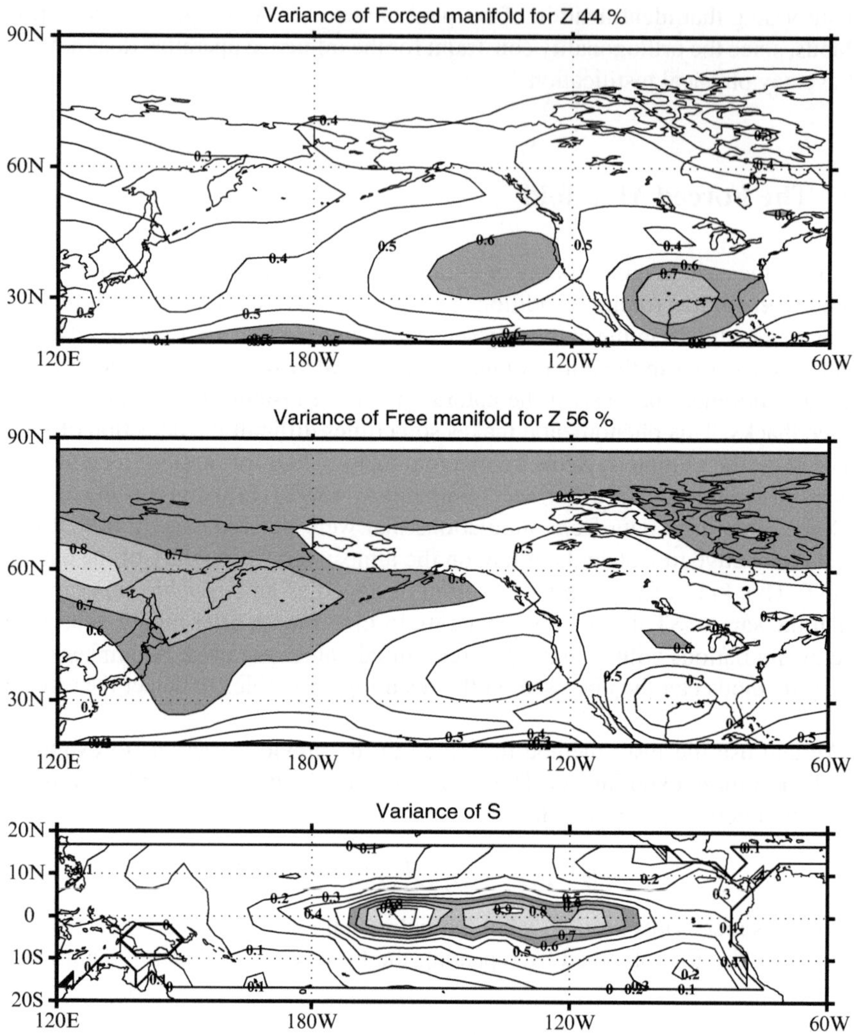

Fig. 8.1 Forced Manifold and Free Manifold. The picture represents the division in Forced and Free Manifold for the North American geopotential field **Z** in the region north of 20N and the tropical Pacific SST in the region 20N–20S latitude. The *top panel* shows the ratio of the variance of the Forced Manifold to the total variance, the *middle panel* shows the same for the Free Manifold. The region used to examine the effect of the SST is in the *bottom panel* together with the total variance of the SST

as the sum of the variance of the Free and Forced Manifold sum to the total variance. There are large differences in the amount of variance in the Forced Manifold from point to point, the differences identify the area where the influence of the SST is felt the most. In this way we can easily identify the region where the variability of some field like the geopotential field is mostly affected by the variance of the other field, in this case the SST.

The Forced and Free manifolds are defined with respect to the regions of the field **S** that we employ as the forcing region. In Fig. 8.2 we have moved the reference SST region to the South Atlantic, more precisely the region between 20S and 70S.

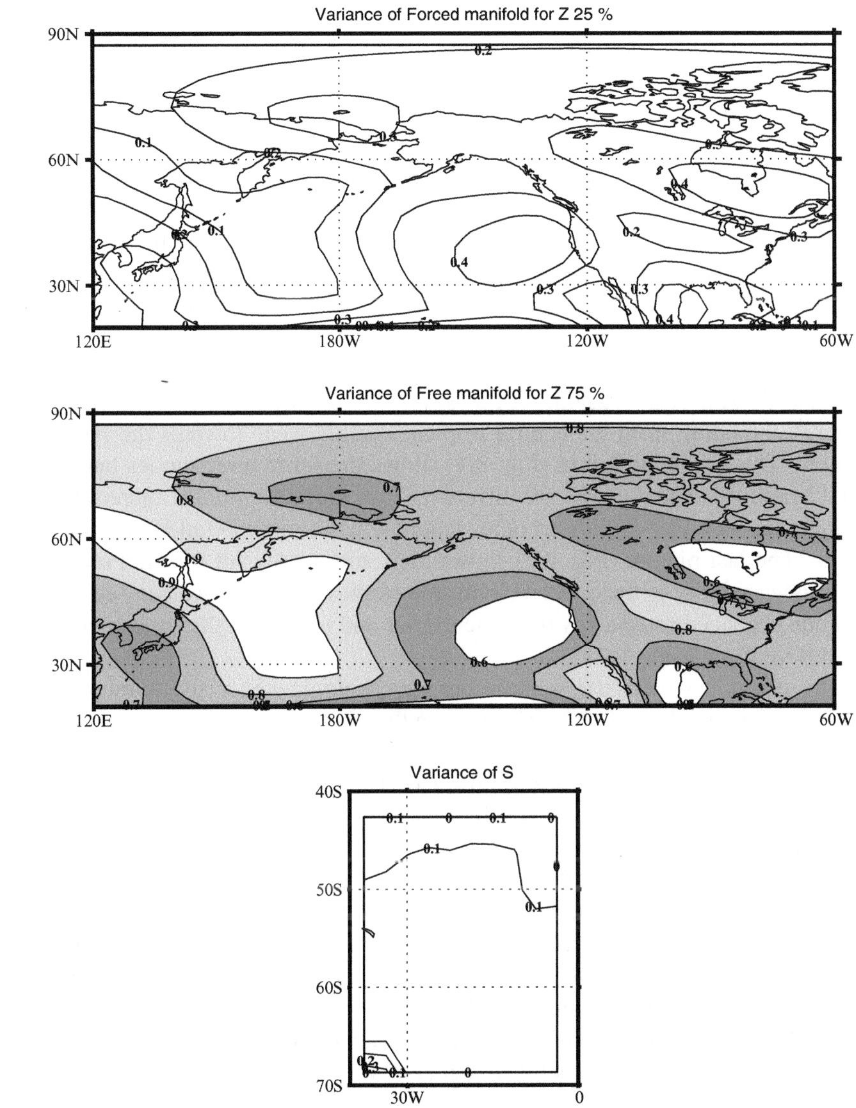

Fig. 8.2 Forced Manifold and Free Manifold. The picture represents the division in Forced and Free Manifold for the North American geopotential field **Z** in the region north of 20N and the Atlantic SST in the region 20S–70S latitude. The *top panel* shows the ratio of the variance of the Forced Manifold to the total variance, the *middle panel* shows the same for the Free Manifold. The region used to examine the effect of the SST is in the *bottom panel* together with the total variance of the SST. The amount of free variance is much larger than in the preceding case, showing the minor impact of the Southern Atlantic SST on the geopotential in the North Pacific

We have physical reasons to believe that the SST in this region are only modestly connected with the activity over the North Pacific. In fact the amount of variance explained by the Forced manifold is much less than before and most of the variance is in the Free Manifold, namely, mostly of the variance is not connected with the variance of the SST in the South Atlantic.

The separation in (8.5) will allow us to study separately the properties of the field that is related to SST from what is independent. We are pretty free to choose the forcing region and the study region any way we wish and so we can select them on the basis of the various scientific hypotheses that we may think interesting. We next see whether there are some basic properties that we can immediately point out. The separation in Forced and Free manifold is not trivial, mostly because of the non linear nature of the pseudoinverse used in the solution of the PRO method. We have therefore no reasons to expect any particular relation between the structure of the manifolds and the total field. The variance structure is unrelated and we can convince ourselves by examining the EOF of the total field compared to those of the two manifolds.

Figure 8.3 displays the first three EOF modes of the geopotential over the Pacific/North American region. These show the familiar positive and negative structures extending from the central tropical Pacific ocean towards the American continent. The following figure (Fig. 8.4) shows the same three modes but for the Forced Manifold. The patterns are different. They show significant increase in the amplitude in the lower tropics ad other features that a more detailed analysis may indicate. The last picture (Fig. 8.5) shows the same modes but for the Free Manifold. This variability is unconnected with the tropical SST and we can see that the amplitude is concentrated away from the tropics and in general the patterns are even more different.

In this example we can use some of our meteorological expertise to interpret the results and connect them to the phenomenon that they represent; in a particular application one will have to do the same, by analyzing in detail the process to discuss the modes in a scientifically meaningful way. It is clear however that the separation is real and that is not trivial, in the sense of simply selecting some particular mode of variations over the others. The main reason for this behavior is the decrease of degrees of freedom due to the pseudoinverse. The pseudoinverse eliminates the degrees of freedom corresponding to zero singular values which are associated to modes that do not contribute to be variance. Hence the Forced Manifold $\mathbf{AS}$ is not the entire space $\mathbf{Z}$. In the case of multiple realizations with the same $\mathbf{S}$, the number of degrees of freedom of the data $\mathbf{Z}$ is not the number of realization times the time levels, in our case 102, but only the smaller number of degrees of freedom corresponding to the time levels for $\mathbf{S}$, namely 34.

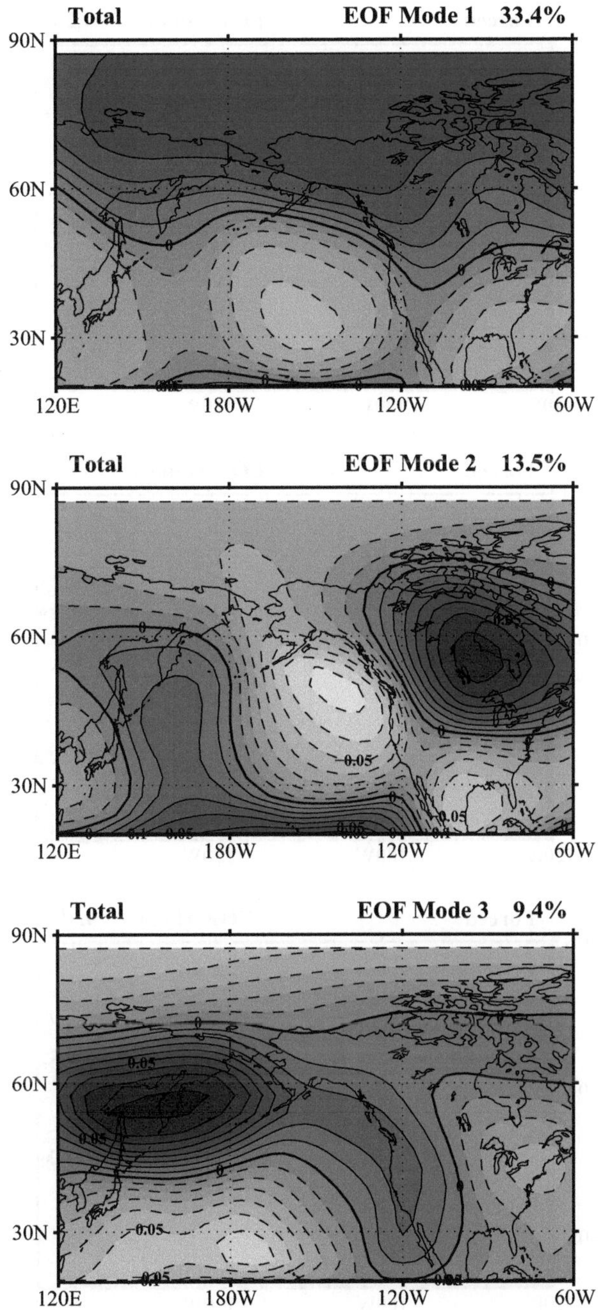

Fig. 8.3 EOF of the Total field. The picture shows the first three EOF modes for **Z** over the region of the Pacific/North America

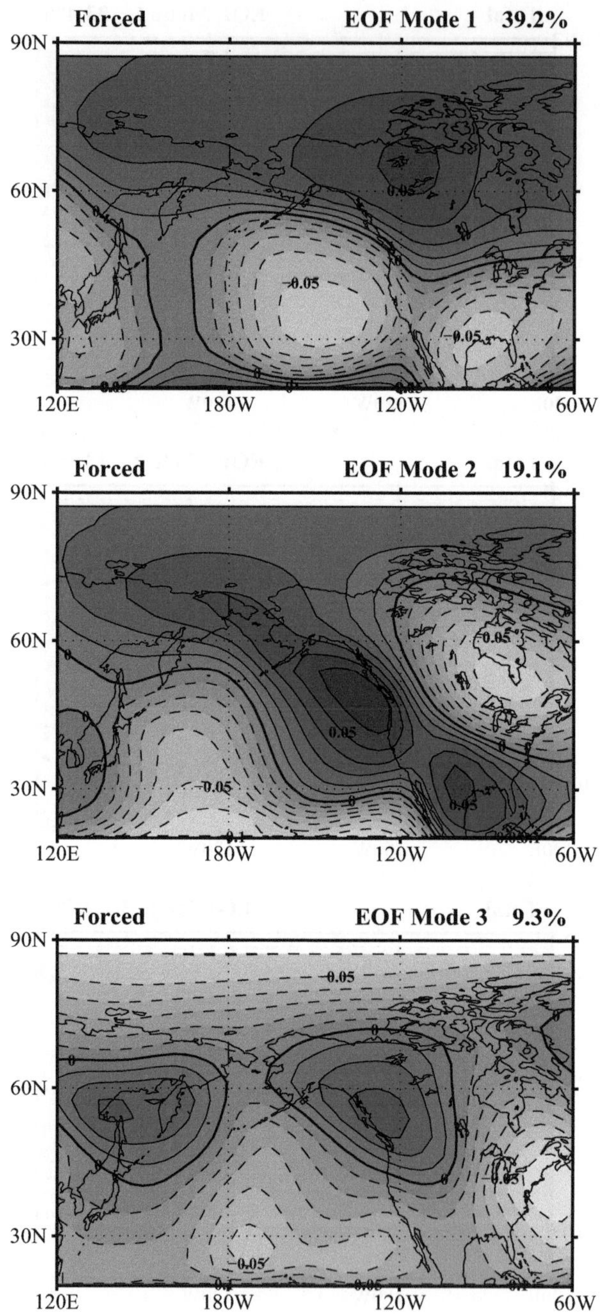

Fig. 8.4 As in Fig. 8.3, but for the Forced Manifold

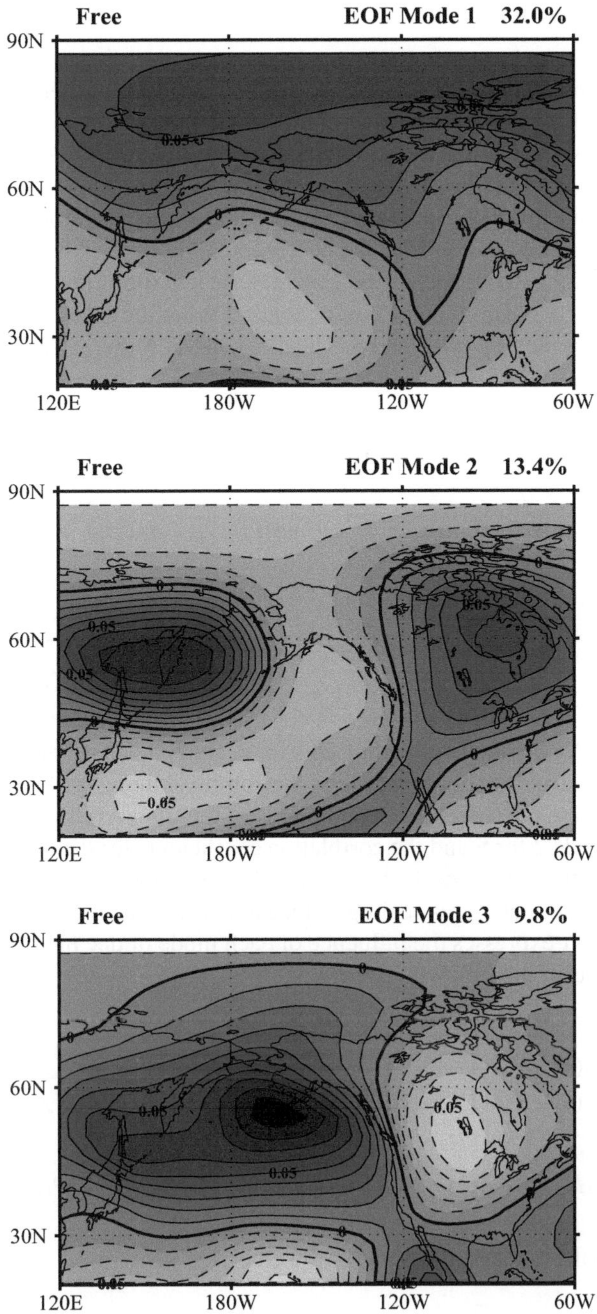

Fig. 8.5 As in Fig. 8.3, but for the Free Manifold

8.3.1 Significance Analysis

The PRO method can give us an easy way to find connections between fields, however there is still the chance that some of the connections are just incidental. It would be nice to have some method to estimate the probability that the separation is indeed the result of casual relations.

To understand this point we need to go back to the operator $\mathbf{A} = \mathbf{Z}\mathbf{S}^T$. The discussion is easier if we consider $\mathbf{A}$ under the scaling of Sect. 8.2.1. This choice is no loss of validity with respect to the general case and the structure of $\mathbf{A}$ is simpler. The definition of $\mathbf{A}$ means that the components of $\mathbf{A}$ are the time correlation between the time series of the EOF of $\mathbf{Z}$ and $\mathbf{S}$. The elements of $\mathbf{A}$ therefore can be written as

$$\begin{bmatrix} a_{11} = \sum_{i=1}^{n} z_1(i)s_1(i) & a_{12} = \sum_{i=1}^{n} z_1(i)s_2(i) & \dots & a_{1N_S} = \sum_{i=1}^{n} z_1(i)s_{N_S}(i) \\ a_{21} = \sum_{i=1}^{n} z_2(i)s_1(i) & a_{22} = \sum_{i=1}^{n} z_2(i)s_2(i) & \dots & a_{2N_S} = \sum_{i=1}^{n} z_2(i)s_{N_S}(i) \\ \vdots & \vdots & \vdots & \vdots \\ a_{N_Z1} = \sum_{i=1}^{n} z_{N_Z}(i)s_1(i) & a_{N_Z2} = \sum_{i=1}^{n} z_{N_Z}(i)s_2(i) & \dots & a_{N_ZN_S} = \sum_{i=1}^{n} z_{N_Z}(i)s_{N_S}(i) \end{bmatrix},$$

where N_Z and N_S are the number of EOFs for $\mathbf{Z}$ and $\mathbf{S}$ that have been retained in the analysis. The matrix is non-symmetric as the components a_{12} and a_{21} are in general different. They are the correlation coefficients of the time series of EOF mode 1 for $\mathbf{Z}$ with EOF mode 2 for $\mathbf{S}$ and the correlation coefficients for the time series of EOF mode 2 of $\mathbf{Z}$ with EOF mode 1 of $\mathbf{S}$. In fact they are indeed regression coefficients in the general case that in this scaling reduce to correlation coefficients. We can see then the matrix $\mathbf{A}$ expresses the influence of each mode of one field on the modes of the other field as in the following scheme

$$\begin{bmatrix} \mathbf{S}(1) \rightarrow \mathbf{Z}(1) & \mathbf{S}(1) \rightarrow \mathbf{Z}(2) & \dots & \mathbf{S}(1) \rightarrow \mathbf{Z}(N_Z) \\ \mathbf{S}(2) \rightarrow \mathbf{Z}(1) & \mathbf{S}(2) \rightarrow \mathbf{Z}(2) & \dots & \mathbf{S}(2) \rightarrow \mathbf{Z}(N_Z) \\ \vdots & \vdots & \vdots & \vdots \\ \mathbf{S}(N_S) \rightarrow \mathbf{Z}(1) & \mathbf{S}(N_S) \rightarrow \mathbf{Z}(2) & \dots & \mathbf{S}(N_S) \rightarrow \mathbf{Z}(N_Z) \end{bmatrix}, \tag{8.9}$$

or introducing the numerical values for the elements

$$\begin{bmatrix} -0.4023 & 0.0126 & 0.2145 & -0.0758 & \dots \\ -0.5741 & 0.2645 & -0.0249 & -0.0797 & \dots \\ 0.0491 & -0.0129 & 0.0787 & 0.1200 & \dots \\ 0.1705 & 0.4186 & 0.0880 & 0.1497 & \dots \\ 0.2999 & 0.1294 & -0.2053 & 0.0469 & \dots \\ \vdots & \vdots & \vdots & \vdots & \vdots \end{bmatrix}, \tag{8.10}$$

where we have shown the first few rows and columns of **A**. We can see that we can use **A** to inspect the strength and the characteristic of the correlation an/or regression between each particular modes and the others. The arguments is not limited by the choice of the representation in EOF. If we had elected to used the grid point or station representation, the operator **A** could have been interpreted in the same way. In that case the elements a_{ij} would have contained the correlation/regression coefficient for grid point i of **Z** with grid point or station j of **S**.

The description in (8.10) and (8.9) indicates that we have a statistical interpretation for **A**. Such an interpretation may be used to establish confidence limits in the numerical vales of the components of **A**. We can used heuristic methods to establish the baseline values that we can attribute to chance. For instance, in the previous sections we have changed the analysis domain to regions where we were expecting varying strengths of the relationship between **S** and **Z**. We can have an idea of the sensitivity of the analysis by also scrambling in time one of the fields and using the method to estimate the possibility of casual relations. The results are shown in Fig. 8.6.

Here we can see that the amount of variance in the Forced Manifold has decreased by a large amount. This level is basically equivalent to the determination of a zero level, that is the value that is generated by casual relations in the data. The amount of Forced variance found is very close to the level determined by the example in Fig. 8.2 where we used a physically based argument to estimate the level of no relation, form the result of the scrambled test we can be rather confident that the relation found in Fig. 8.1 is relatively robust.

We can have a more rigorous estimation of confidence if we recognize that the correlation/regression coefficients in **A** can be tested against a Student's t-distribution. The test estimates the probability that the true coefficient is indeed zero. The acceptable values for the probability levels in order to accept the computed values is, of course, matter of choice, but usually values of 5% or 1% are used. These choices correspond to the statement that there a 5% or 1% probability that the hypothesis that the true value is indeed zero is true. We can insert this process into the calculation of the Forced Manifold by testing each element of **A** and putting to zero those components that pass the test. We can repeat the calculation for Fig. 8.1 introducing now the significance test at 1%. The results are shown in Fig. 8.7.

The Forced and Free manifolds for the geopotential and the tropical SST have a similar distribution as in Fig. 8.7. Overall the total amount of variance that can be attributed with confidence to the forcing **S** is decreased, but the distribution has concentrated and the difference between maxima and minima has increased. The

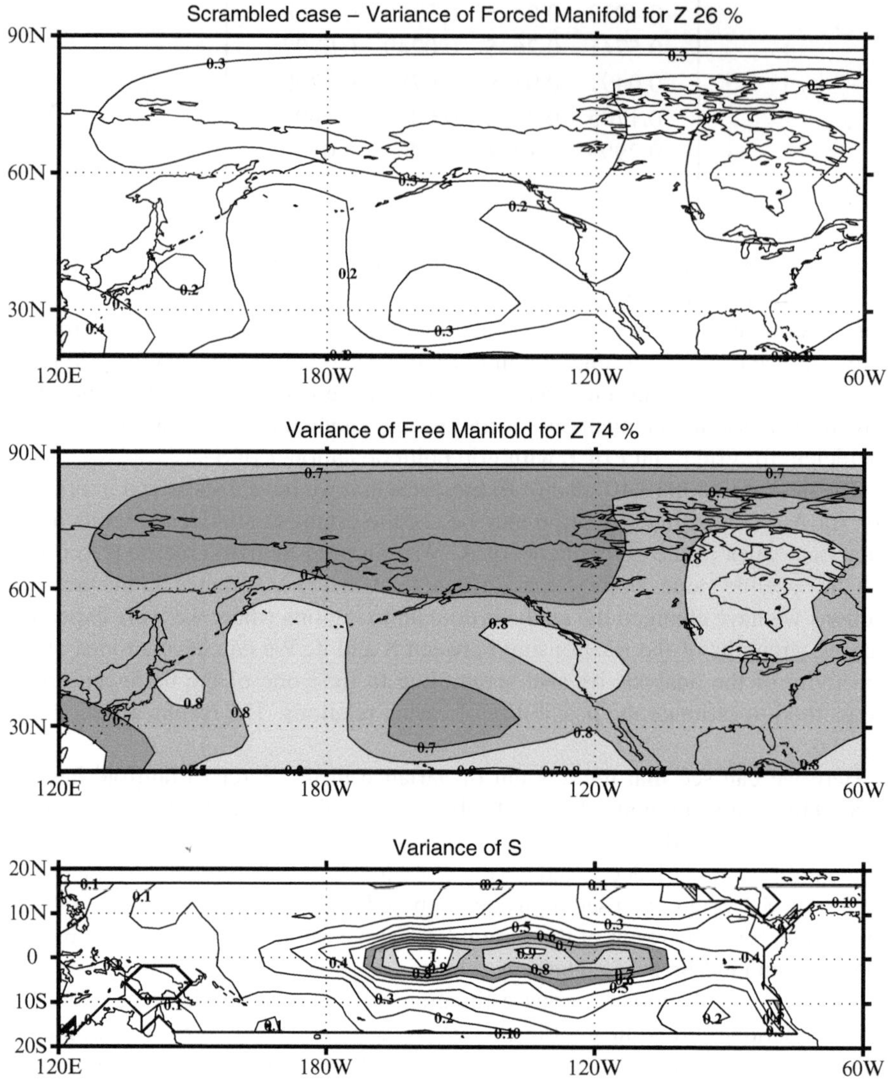

Fig. 8.6 As in Fig. 8.1, but for a randomly permuted field in time. The amount of variance that such casual relations detect is much lower than in the original picture, giving us some confidence in the determination of the original **A**. The amount of variance found is also very close to the physically motivated examples of Fig. 8.2, adding more confidence to the robustness of the original estimation

coefficients of the operator **A** that are significant according to the Student *t*-test are indicated in the top right panel and they have been retained in the calculation, whereas the other have been put to zero. The significance test makes it easy to deal with the time scrambled case (Fig. 8.8). There is no coherent pattern in the **Z** field and the relations seem completely casual.

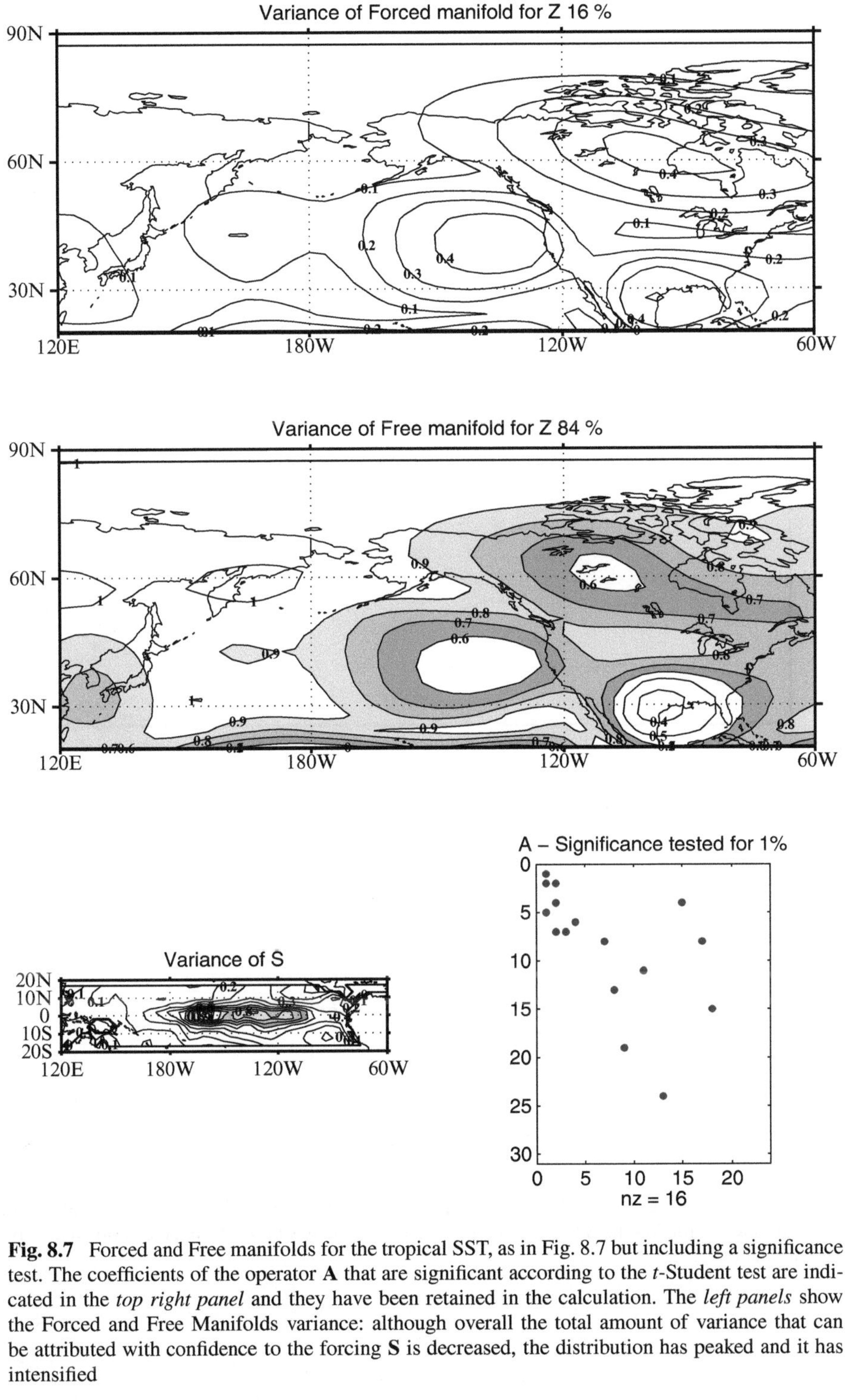

Fig. 8.7 Forced and Free manifolds for the tropical SST, as in Fig. 8.7 but including a significance test. The coefficients of the operator **A** that are significant according to the *t*-Student test are indicated in the *top right panel* and they have been retained in the calculation. The *left panels* show the Forced and Free Manifolds variance: although overall the total amount of variance that can be attributed with confidence to the forcing **S** is decreased, the distribution has peaked and it has intensified

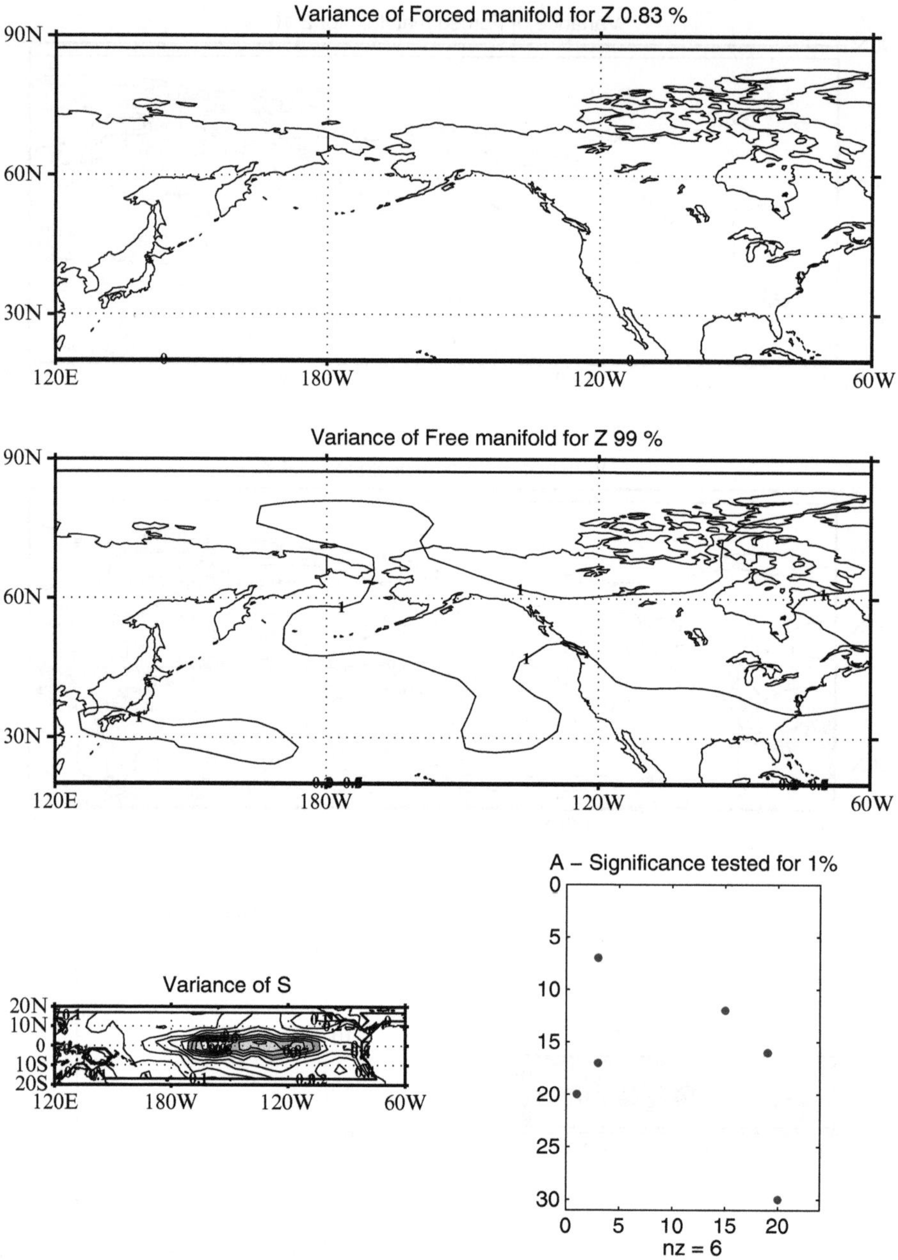

Fig. 8.8 Forced and Free manifolds for the tropical SST, as in Fig. 8.7, including a significance test and time scrambling the **S** field. No coherent pattern emerges in the **Z** field indicating a mostly casual relation

8.4 The Coupled Manifold

The analysis of the ensemble experiments with multiple realization showed that the PRO method is very efficient at identifying the influence of one field on the other, but this is a somewhat easy case. The presence of multiple cases with the same forcing, the so-called ensemble, makes it possible to apply other techniques for identifying the portion of variance that is linked to the external forcing itself. Methods like separation of variance (Rowell 1997) or even the simple usage of the ensemble mean can give a good indication of the characteristics of the forced response, even in case they miss the detailed separation of the time series of the field themselves. All these approaches however fail when we are confronted with the case of a single realization.

There are several cases in which executing ensembles is impossible or forbidden by the terms of the physical problem. In general this can happen when the two fields that we want to examine are part of the same dynamically linked problem and they cannot be separated in a "forcing" and a "response". Coupled atmosphere–ocean climate simulations are in this class with regard to our examples of marine temperatures (SST) and atmospheric geopotential $\mathbf{Z}$, since in this case the evolution of the SST is not prescribed externally but is partially determined by the geopotential itself. We then have to investigate to what extent the geopotential exerts control over the SST.

Another notable example are observation records that are not reproducible. In many cases experiments can be repeated and statistical ensembles can be constructed but in geophysical application observations cannot be reproduced in a strict sense. The Earth atmosphere and ocean constantly evolve and our record of observations in time is a single realization of the Earth climate. The situation is very similar to a numerical simulation performed with a coupled atmosphere–ocean model, also in this case no parameter can be considered "external" and traditional separation of variance methods fail.

However the formulation of the Forced Manifold is sufficiently general that it can be used also in the case in which we have a single realization. Nothing in the formulation we have used in (8.3) or (8.8) is linked to the availability of multiple realizations. We can set up the problem also for single data sets $\mathbf{Z}$ and $\mathbf{S}$. The only victim is probably the name, since in this case we do not have "forcing" field and "response" field and calling it "Forced Manifold" does not seem very appropriate. We can still separate the field in sectors, but now we have a mutual effect of one field on the other, then the name *Coupled Manifold* rather Forced Manifold seems more appropriate. The Free Manifold, instead, maintains its meaning of variance that is free from the influence of the field under examination.

The results are shown in Fig. 8.9. We present here the results obtained both under the problem $\mathbf{Z} = \mathbf{AS}$ and $\mathbf{S} = \mathbf{BZ}$. The top line shows the Coupled Manifold for $\mathbf{Z}$ and the corresponding Free Manifold. We can see a familiar pattern of locations where the variance of $\mathbf{Z}$ is highly influenced by the variations of the SST in the region. The non-local nature of the analysis means that we can conclude only that the various geographical locations in $\mathbf{Z}$ are globally influenced by the entire region

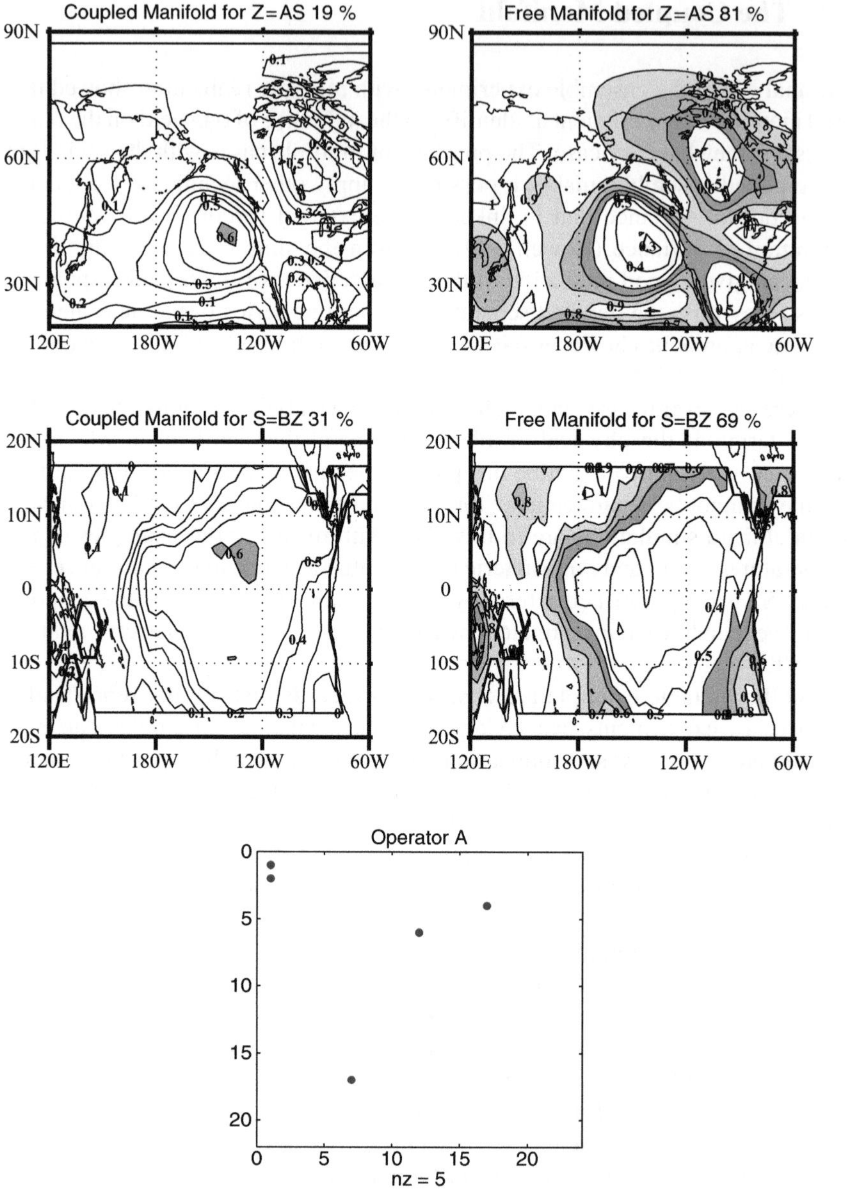

Fig. 8.9 Coupled and Free manifolds for the tropical SST and the North America geopotential. The significance testing is included. The picture shows the result for the problem $\mathbf{Z} = \mathbf{AS}$, *top row*, and for the problem $\mathbf{S} = \mathbf{BZ}$, *bottom row*. The significant element of the operator $\mathbf{A}$ that is used in the two problems are shown. Remember that $\mathbf{B} = \mathbf{A}^T$ in this calculations. The areas with a ratio larger than 0.6 are *shaded*

in **S**. The other problem $\mathbf{S} = \mathbf{BZ}$ gives us the opportunity to see the geographical distribution of the influence of the entire **Z** region over the SST variability. We are looking here at distant regions that show very well the flexibility of these methods: they can be applied to varying field with few limitations in space and time, but they must be interpreted with caution.

These methods identify patterns of co-variation that may or may not correspond to physical causal relation between the fields. For instance, we have theoretical arguments to expect the influence by the tropical SST on the geopotential in the North America sector and the methods very nicely allow us to investigate this relation in detail. The opposite formulation makes it possible to investigate instead the influence of the geopotential **Z** on the tropical SST and to inspect its geographical distribution. Unfortunately, in this case we do not have a theory for the influence of the American geopotential on the tropical SST, so what we see here is basically a representation of the co-variation relations observed in the previous $\mathbf{Z} = \mathbf{AS}$ case. These methods cannot really provide us with causal relation, but they can point us to the right direction. Only our scientific ingenuity and arguments can then transform them into cause–effect theories.

The situation may change if we analyze fields where we know on physical grounds that some interaction is present and therefore some mutual influence exists. We will expand here slightly our data base extending the test case to a time series of monthly means of SST and the east–west component of the surface winds, expressed here as wind stress. The data have been obtained from a long simulation with a coupled model at low resolution and extended for 200 years.

The results shown in Fig. 8.10 indicate a different result from the preceding one. We know that SST and this component of the surface Wind are strongly interacting in this region, and we have selected the same geographical domain for the two fields. The distribution of the Coupled Manifold is consistent with each other as we may expect in a situation where the two fields exert a mutual influence on each other. The fraction of variance in the Coupled Manifold is locally very high and values of 70% are reached. The joint variability region is mostly limited to the equatorial Pacific and it becomes weaker moving to the higher latitudes.

A random permutation of the time series of SST (Fig. 8.11) completely destroys the relation. The Coupled manifold disappears except for a small residue. The effect is more dramatic than in the previous scrambling example using SST and the geopotential; the larger response is due to the stronger couplings that exist between the SST and the wind and also to the longer time series. The latter is what makes the relation between the fields more accurate.

These results are quite general and they are not limited to this atmosphere-ocean example. In general longer time series will improve the accuracy and the reliability of the identification of the relation.

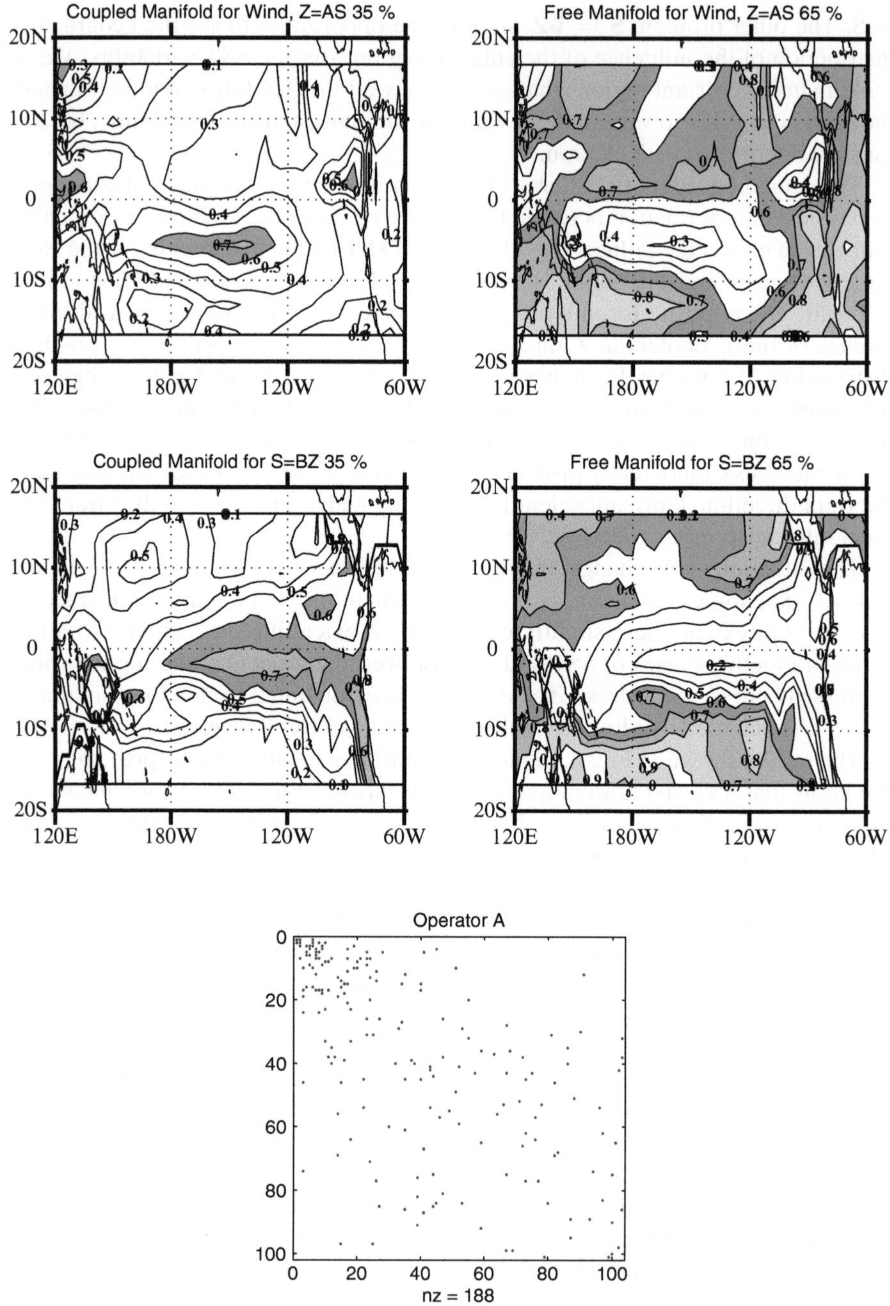

Fig. 8.10 Coupled and Free manifolds for the tropical SST and surface east–west Wind in the same region. Significance testing at 1% is included. The picture shows the result for the problem $\mathbf{Z} = \mathbf{AS}$, *top row*, and for the problem $\mathbf{S} = \mathbf{BZ}$, *bottom row*. The significant element of the operator $\mathbf{A}$ that is used in the two problems are shown. The areas with a ratio larger than 0.6 are *shaded*

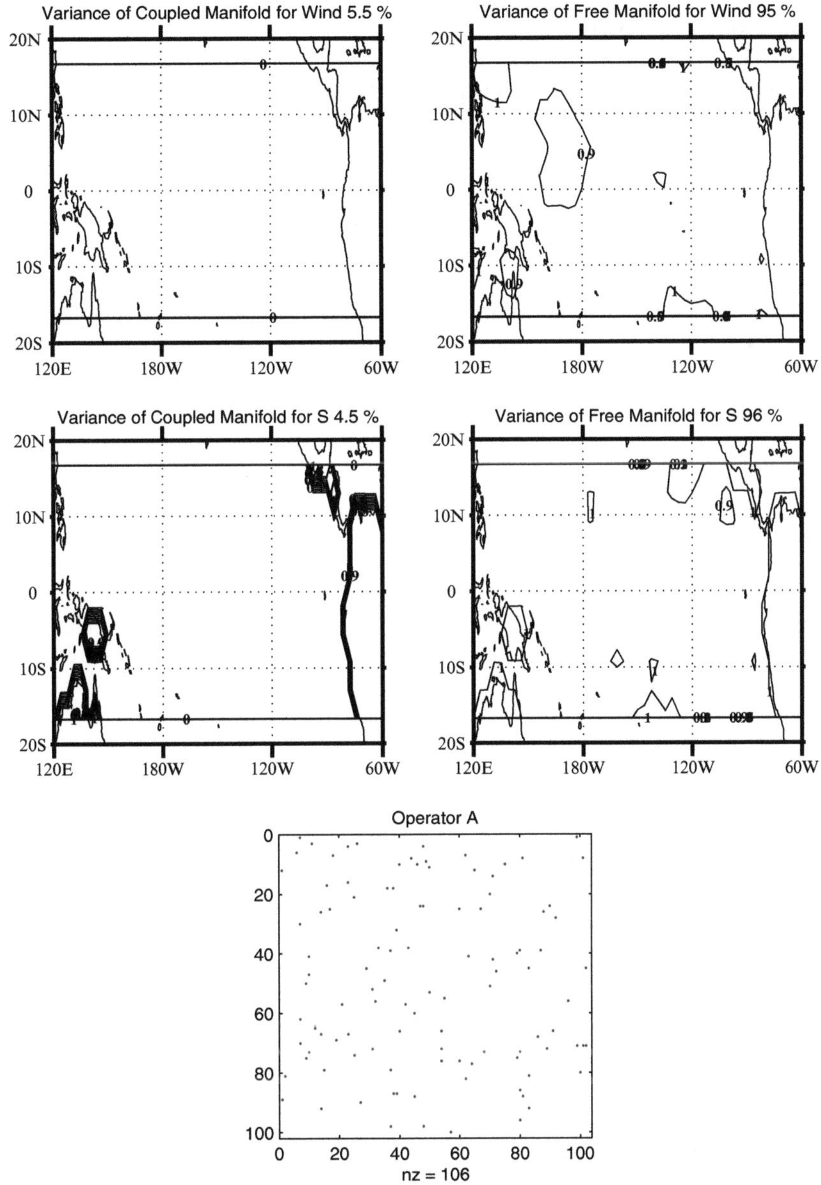

Fig. 8.11 Coupled and Free manifolds for the tropical SST and surface east–west Wind in the same region with a time scrambling of the SST time series. The significance testing is included. The picture shows the result for the problem $\mathbf{Z} = \mathbf{AS}$, *top row*, and for the problem $\mathbf{S} = \mathbf{BZ}$, *bottom row*. The significant element of the operator $\mathbf{A}$ that is used in the two problems are shown. The areas with a ratio larger than 0.6 are *shaded*

Exercises and Problems

1. Using the data sets and the scripts provided compute the EOF of $\mathbf{Z}_S$ and $\mathbf{Z}_{free}$ and check that they are different.
2. Show that the elements of **A** are regression coefficients and that they coincide with correlation coefficients if the variance scaling is used.
3. Try to compute the Forced and Coupled Manifolds for different regions.
4. Modify the significance settings in the calculation and check how the ratio between the Forced and Free parts is modified.
5. Show that each field can be expressed as the sum of three terms, using the operators **AB**.

References

Barnett TP, Preisendorfer RW (1987) Origins and levels of monthly and seasonal forecast skill for united states surface air temperatures determined by canonical correlation analysis. Mon Wea Rev 115:1825–1850

Bretherton CS, Smith C, Wallace JM (1992) An intercomparison of methods for finding coupled patterns in climate data. J Climate 5:541–560

Cherry S (1996) Singular value decomposition and canonical correlation. J Climate 9:2003–2009

Cherry S (1997) Some comments on the singular value decomposition. J Climate 10:1759–1761

Clarke GM, Cooke D (1998) A basic course in statistics, 4th edn. Arnold, London and New York

Golub H Gene, Charles F Van Loan (1996) Matrix computations, 3rd edn. The John Hopkins University Press, Baltimore

Hahn SL (1996) Hilbert transforms in signal processing. Artech House, Norwood, Maryland

Harman HH (1976) Modern factor analysis, 3rd edn. University of Chicago Press, Chicago

Horn RA, Johnson CR (1991) Topics in matrix analysis. Cambridge University Press, Cambridge

Jolliffe IT (2002) Principal component analysis, 2nd edn. Springer Series in Statistics, Springer, New York

matlab7 (September 2004) *MATLAB 7*. The MathWorks, Inc

Meyer CD (2000) Matrix analysis and applied linear algebra. SIAM, Philadelphia

Navarra A, Tribbia J (2005) The coupled manifold. J Atmos Sci 62:310–330

North GR, Bell TL, Cahalan RF (1982) Sampling errors in the estimation of empirical orthogonal functions. Mon Wea Rev 110:699–706

Preisendorfer RW (1988) Principal component analysis in meteorology and oceanography, vol 17, Development in atmospheric sciences. Elsevier, Amsterdam

Roberta Q, Christopher SB, John M Wallace (September 2005) On sampling errors in empirical orthogonal functions. J Climate 18(17):3704–3710 http://dx.doi.org/10.1175/%2FJCLI3500.1

Richman MB, Vermette SJ (1993) The use of procrustes target analysis to discriminate dominant source regions of fine sulfur in the western USA. Atm Environ 27A:475–481

Rowell DP (1997) Using an ensemble of multi-decadal gcm simulations to assess potential seasonal predictability. J Climate 11:109–120

von Storch H, Zwiers FW (1999) Statistical analysis in climate research. Cambridge University Press, Cambridge

Wilks S Daniel (2005) Statistical methods in the atmospheric sciences, 2nd edn. International Geophysics series, Academic Press, Oxford

Index

FSC
www.fsc.org
MIX
Papier aus verantwortungsvollen Quellen
Paper from responsible sources
FSC® C105338

If you have any concerns about our products,
you can contact us on
ProductSafety@springernature.com

In case Publisher is established outside the EU,
the EU authorized representative is:
Springer Nature Customer Service Center GmbH
Europaplatz 3, 69115 Heidelberg, Germany

Printed by Libri Plureos GmbH
in Hamburg, Germany